Locating Lines and Hyperplanes

Applied Optimization

Volume 25

Series Editors:

Panos M. Pardalos
University of Florida, U.S.A.

Donald Hearn
University of Florida, U.S.A.

The titles published in this series are listed at the end of this volume.

Locating Lines and Hyperplanes

Theory and Algorithms

by

Anita Schöbel

Universität Kaiserslautern,
Kaiserslautern, Germany

SPRINGER-SCIENCE+BUSINESS MEDIA, B.V.

A C.I.P. Catalogue record for this book is available from the Library of Congress.

ISBN 978-1-4613-7428-2 ISBN 978-1-4615-5321-2 (eBook)
DOI 10.1007/978-1-4615-5321-2

Printed on acid-free paper

For my children
Svenja and Malte

Contents

A straight line is one which lies evenly with the points on itself.

—Euclid, The Elements

Straight lines have been a fascinating subject in mathematics right from the beginning of geometry.
But even in modern mathematical disciplines affine subspaces like lines and hyperplanes are still an exciting research object. In this text I am concerned with the location of lines in the plane and of hyperplanes in normed spaces.

On a conference on transportation analysis I first heard of path location problems in networks. Such problems deal with the location of a path which is as close as possible to a given set of customer-nodes. In **traffic planning** these problems arise when a new bus line is organized, or if a new railway line is built. Since Professor Hamacher and my colleagues work on *continuous* location problems I decided to study path location problems, not in networks, but in the plane. Unfortunately, the location of an arbitrary or of a polygonal path is NP-hard (see Section 8.3) such that I turned my attention to the location of straight lines.
Given a set of weighted points in the plane, the line location problem is to find a straight line which minimizes the sum of distances or the maximum distance to the given point set. Studying the literature about the location of lines, I recognized that these problems play an important role not only in **location theory**, but also in **statistics** when finding regression lines (*orthogonal or vertical L_1-fit problem*) and in **computational geometry** as a special case of hyperplane approximation problems (*L_1- or L_∞ approximation problems*).
In this text, I present my research on line and hyperplane location problems. Figure 0.1 shows the relations between the chapters of this text. A brief outline of all chapters is given in the following:

- In Chapter 1, I formally introduce hyperplane location problems with respect to different distance measures. I also discuss the relation of these problems to transversal theory and to the set width problem in computa-

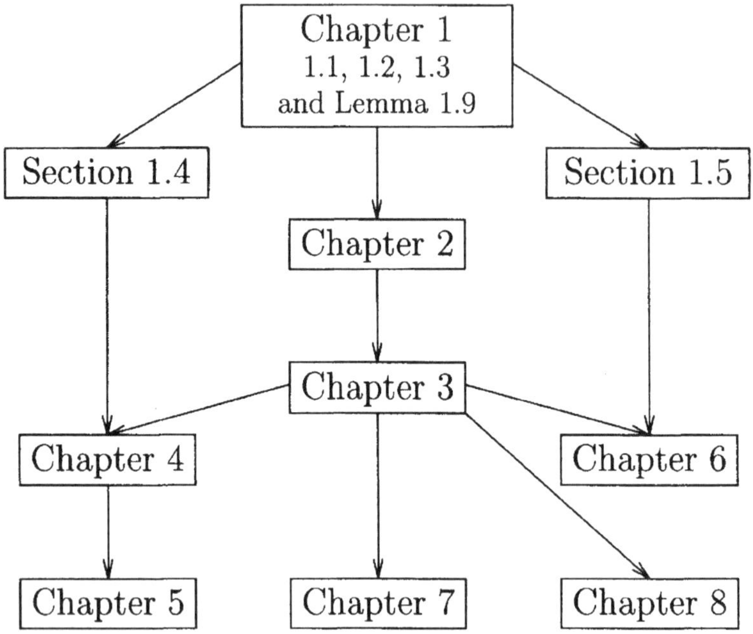

Figure 0.1. Relations between the chapters of this text.

tional geometry (Section 1.4). In Section 1.5, I introduce the concept of restricted piecewise linear convex programs which is needed in Chapter 6.

- Chapter 2 deals with line location problems with the vertical distance. Since these problems are convex, they can easily be solved. In this chapter, I also introduce a geometric duality which gives some new insight in line location problems with vertical distance. Chapter 2 is an important building block for all subsequent chapters.

- Then I turn to (non-convex) line location problems with arbitrary norms (Chapter 3). After reviewing the known results about line location problems with rectangular and Euclidean distance, I extend the results of Chapter 2 to all distances derived from norms. Algorithmic approaches for arbitrary norms and for the case of block norm distances are given.

- In Chapter 4 I can strengthen the results of Chapter 3 in the case of smooth norms. These results lead to algorithms that find the set of *all* optimal lines, even in the case of arbitrary norms.

- Chapter 5 investigates distance functions that are not derived from norms. In particular, I study the case of gauges, of metrics, and of mixed distance functions.

- Suppose that we have a given forbidden region in the plane that must not be intersected by the new line facility. Such restricted line location problems are discussed in Chapter 6. First, I consider the case of the vertical distance and then extend the derived results to block norms and arbitrary norms. Additionally, I show that the location of a line segment with fixed length can be modelled as a restricted line location problem in the case of the vertical distance. Finally, some other types of restrictions will be discussed.

- The location of hyperplanes is the subject of Chapter 7. Here I extend the concepts developed in Chapters 2, 3, and 4 to higher dimensions. First, a dual interpretation of hyperplane location problems will be given, then I deal with the location of hyperplanes with respect to arbitrary norms. Algorithmic approaches which can also be used for high-dimensional stabbing problems follow, and finally I present a characterization of smooth norms.

- In Chapter 8 the location of objects other than affine subspaces is discussed. I first deal with the bicriteria problem of locating line segments, minimizing the length of the segment and the distance to the existing facilities. Then some remarks about the location of a circle in the plane are added, and in the last section I give a general formulation of the location of one-dimensional facilities and show relations to well-known optimization problems.

Last, but not least I take the opportunity to say "Thank you!" to all people who supported me while writing this book. First of all, I want to thank Professor Horst W. Hamacher for numerous fruitful discussions, his interest in my work, and for the pleasure it was to be a member of his working group. He made it possible for me and encouraged me to do both, work in mathematics and care for my children.
Second, I want to express my thanks to Professor Horst Martini for turning my attention to hyperplane location problems, for many valuable comments and for our joint work which resulted in several papers.
My thanks also go to Jörg Fliege, Stefan Nickel, Frank Plastria, and Margaret Wiecek for their cooperation and various helpful ideas. I am grateful to Sharon Donovan and John Martindale at Kluwer for their friendly support.
I am much obliged to my colleagues Matthias Ehrgott, Kathrin Klamroth, and Martin C. Müller, who read the manuscript of this work, tried hard to improve my English and provided helpful suggestions. To Martin I want to express some special thanks for his kind and competent assistance in LaTeX.

My most particular thanks go to my husband Georg whose understanding and encouragement made it possible for me to do this research.

Anita Schöbel Kaiserslautern, October 1998

1
BASIC CONCEPTS

1.1 Introduction

Before we discuss line location problems in the plane and the location of hyperplanes in $\mathbb{R}^n$, we will give some insight where such problems can arise and try to give a feeling for the mathematical fields in which line location plays an important role.

Location theory: In location theory we are given a set of existing facilities in the plane with nonnegative weights representing the importance of the existing facilities. We want to locate one or more new facilities as close as possible to the given set of existing facilities. Usually, as objective function, the average (weighted) distance or the maximum (weighted) distance between the existing facilities and the new facilities has to be minimized. A lot of extensions are possible. One example is the introduction of any type of restriction which can be given by a forbidden area where no new facility is allowed to be placed. Other extensions include the discussion of negative weights in obnoxious facility location or the introduction of any special distance measure between the existing facilities and the new ones. For an overview about classical facility location in the plane, see e.g. [Ham95, Pla95, LMW88].

Line location can be seen as an extension of classical facility location problems. Instead of a single point we want to locate a line in the plane. The objective function is the same as in classical facility location, namely to minimize the distances (weighted average distance or weighted maximum distance) between a given set of existing facilities and the new line. From that point of view, line location problems belong to the area of path location, i.e., the location of dimensional facilities such as lines, line segments, trees, or paths. These problems have been extensively discussed not in the plane, but in networks, for an overview see [MB96] or [HSL93]. A recent survey about the location of dimensional structures in the plane can be found in [Mes95].

Possible applications include the planning of new railways or motorways, where the existing facilities may be cities and the weights represent the number of their inhabitants. Path location can also be used to determine the location of pipelines, drainage or irrigation ditches or in the field of plant layout, see for example [MN80].

Computational Geometry: Line location problems are also known in computational geometry. Here they are called *linear L_1 or L_∞ approximation problems.* The denotation L_1 refers to the minimization of the average distance and L_∞ refers to the maximum distance between the given points and the line. For an overview of these problems in computational geometry see e.g. the surveys [Lee86, LW86, Kor89, KM93]. In particular, the time complexity of line location problems was investigated by several authors, cf. [MT83, YKII88, HIIR89, KM90, KM93] and [HII+93].

Statistics: Another field in which line location problems play an important role is robust statistics (and also in numerical mathematics) where these problems are known as absolute errors regression, median problems, L_1 regression and orthogonal/vertical L_1–fit problems , respectively. In fact, finding a regression line is nothing else than a line location problem with some special function to measure the distance between the given points and the regression line. Problems of this type are investigated in [Bos57, Edg88, Ric64, PFTV86] and [SW87].

Since at least three mathematical disciplines are involved in line location problems, the literature about such problems is not very homogeneous. Different proofs for the same theorem can be found and often these proofs use quite different approaches (compare the proof of Lemma 3.4 or the algorithmic approaches mentioned in Section 2.1). In the following chapters we will use both methods of classical facility location and geometric properties to discuss line location problems. Our main purpose is to discuss line location problems with other distance measures than the Euclidean or the rectangular norm. So far, nothing seems to be known about such line location problems in all three mentioned disciplines. A few proofs will be done twice using two different approaches (Lemma 3.6 and Theorem 3.3).

Before we continue, we briefly describe line location problems in the plane in the following way: Given a distance function d and a set of existing facilities in the plane $\mathbb{R}^2$,

$$\mathcal{E}x = \{Ex_1, Ex_2, \ldots, Ex_M\}, Ex_m = (a_{m1}, a_{m2}) \text{ for all } m = 1, \ldots, M$$

with positive weights $w_m > 0$ for all $m = 1, \ldots, M$, we are looking for a straight line l such that we minimize one of the following objective functions:

Median function

$$f(l) := \sum_{m \in \mathcal{M}} w_m d(Ex_m, l),$$

Center function

$$g(l) := \max_{m \in \mathcal{M}} w_m d(Ex_m, l).$$

The distance $d(Ex_m, l)$ from an existing facility Ex_m to the line l is given by the distance from Ex_m to its nearest point on the line, i.e.,

$$d(x, l) = \inf_{y \in l} d(x, y).$$

A more formal description of line location problems will be given in Section 1.3 after the introduction of the used distance measures d.

1.2 Measuring Distances: Norms and Metrics

For modelling location problems it is very important to choose a suitable distance function to measure how far it is to travel from the new facility to the existing ones. This is also necessary for line location problems. In this section we briefly introduce the distance measures which we will investigate in the following chapters. As geometrical observations play an important role for proving some of our main results, it makes sense to give the following geometrical definition of the norms used.

Definition 1 ([Min67]) *Let B be a compact convex set in $\mathbb{R}^n$ with nonempty interior which is symmetric with respect to the origin. Let $x \in \mathbb{R}^n$. Then define the* **norm** $\gamma : \mathbb{R}^n \to \mathbb{R}$ *as*

$$\gamma(x) := \inf\{\lambda > 0 : x \in \lambda B\}.$$

The following Lemma 1.1 says that γ satisfies the properties required for norms, i.e., for all $x, y \in \mathbb{R}^n$ and $\lambda \in \mathbb{R}$ we have

$$\gamma(x) \geq 0 \tag{1.1}$$
$$\gamma(x) = 0 \iff x = 0 \tag{1.2}$$
$$\gamma(\lambda x) = |\lambda|\gamma(x) \text{ and} \tag{1.3}$$
$$\gamma(x + y) \leq \gamma(x) + \gamma(y). \tag{1.4}$$

On the other hand, all norms can be characterized by their **unit balls** B.

Lemma 1.1 ([Min67]) *The following hold:*

1. *Let γ be given as in Definition 1. Then γ satisfies 1.1 - 1.4.*
2. *Let $\gamma : \mathbb{R}^n \to \mathbb{R}_{\geq 0}$ be given, such that γ satisfies 1.1 - 1.4. Then its unit ball $B_\gamma = \{x \in \mathbb{R}^n : \gamma(x) \leq 1\}$ is a compact convex set with nonempty interior which is symmetric with respect to the origin.*

If we drop the assumption that the set B is symmetric we get a gauge function $\tilde{\gamma}$. In this case $\tilde{\gamma}(-x) = \gamma(x)$ does not hold in general and $\tilde{\gamma}$ does not define a norm.

Definition 2 ([Min67]) *Let B be a compact convex set in $\mathbb{R}^n$ containing the origin in its interior. The* **gauge** *of x with respect to B is then defined as*

$$\tilde{\gamma}(x) := \inf\{\lambda > 0 : x \in \lambda B\}.$$

Definition 3 ([WW85]) *If B is a convex polytope the resulting gauge $\tilde{\gamma}_B$ is a* **polyhedral gauge** *, and if, in addition, B is symmetric the resulting norm γ_B is a* **block norm** *. The extreme points $Ext(B) := \{b_1, b_2, \ldots, b_H\}$ of B are also called the* **fundamental directions** *of the polyhedral gauge .*

For block norms we can write (after renumbering and with $G = \frac{H}{2}$)

$$Ext(B) = \{b_1, b_2, \ldots, b_G, -b_1, -b_2, \ldots, -b_G\}.$$

A block norm γ can then be reformulated as

$$\gamma_B(x) = \min\left\{\sum_{g=1}^{G} |\lambda_g| : x = \sum_{g=1}^{G} \lambda_g b_g\right\},$$

see e.g. [Nic95]. The importance of block norms and gauges is established by the following theorem.

Theorem 1.1 ([WW85]) *The class of polyhedral gauges is dense in the set of all gauges, and in particular, the class of block norms is dense in the set of all norms.*

The distance between two points can now be defined as usual by

$$d(x, y) = \gamma(y - x).$$

All distances d which are derived from norms are **metrics** i.e.,

$$\begin{aligned} &d(x, y) \geq 0 \\ &d(x, y) = 0 \iff x = y \\ &d(x, y) = d(y, x) \\ &d(x, z) \leq d(x, y) + d(y, z). \end{aligned}$$

Additionally they are **translation invariant** , i.e., the sets

$$\begin{aligned} B(x) &:= x + B \\ &:= \{x + y : y \in B\} \\ &= \{y \in \mathbb{R}^n : d(x,y) \leq 1\} \end{aligned}$$

are translates of the unit ball B for all $x \in \mathbb{R}^n$.
If the distance d is derived from a gauge $\tilde{\gamma}$ then it is still translation invariant, but no longer a metric, as $d(x,y) = d(y,x)$ does not hold in general.
In the following we repeat the definition of the most common family of norms.

Definition 4 (see e.g. [For83]) *Let $x \in \mathbb{R}^n$. Then for $p \in [1,\infty)$ the* **p-norm** *of x is defined by*

$$l_p(x) = \sqrt[p]{\sum_{i=1}^{n} |x_i|^p}$$

and for $p = \infty$ by

$$l_\infty(x) = \max_{i=1,\ldots,n} |x_i|.$$

For $x, y \in \mathbb{R}^n$ the **p-norm distance** between x and y is therefore given as

$$l_p(x,y) = \begin{cases} \sqrt[p]{\sum_{i=1}^{n} |x_i - y_i|^p} & \text{for } p \in [1,\infty) \\ \max_{i=1\ldots n} |x_i - y_i| & \text{for } p = \infty. \end{cases}$$

Some remarks about the p-norms will be added.

For $p = 2$ we get the **Euclidean distance** distance. The **Euclidean length** of a vector $t \in \mathbb{R}^2$ is given by $l_2(0,t)$ and will be denoted by $\|t\|$.

The norm l_p is a block norm if and only if $p \in \{1,\infty\}$. The special case l_1 is called **Manhattan** distance or **rectangular** distance. Its unit ball in $\mathbb{R}^n$ is given by the extreme points

$$Ext(B_{l_1}) = \{e_i, -e_i : i = 1, 2, \ldots, n\},$$

where $\{e_1, e_2, \ldots, e_n\}$ denotes the set of unit vectors of $\mathbb{R}^n$.
For $p = \infty$ we get the l_∞ norm, called **Chebyshev** norm, **maximum** norm, or **one-infinity** norm. It is given by

$$l_\infty(x) = \max\{|x_1|, |x_2|, \ldots, |x_n|\}$$

with the following extreme points of its unit ball:

$$Ext(B_{l_\infty}) = \{(s_1, s_2, \ldots, s_n) : s_i \in \{1, -1\}\}.$$

Note that for two dimensions there exists a linear and norm-converting mapping $T_{l_1,l_\infty} : \mathbb{R}^2 \to \mathbb{R}^2$ between l_1 and l_∞:

Lemma 1.2 ([FMW92, Ham95]) *For the transformation T_{l_1,l_∞} given by*

$$T_{l_1,l_\infty} = \frac{1}{2}\begin{pmatrix} 1 & 1 \\ -1 & 1 \end{pmatrix}$$

the following hold:

1. $l_\infty(x,y) = l_1(T_{l_1,l_\infty}(x), T_{l_1,l_\infty}(y))$ *and*
2. $l_1(x,y) = l_\infty(T^{-1}_{l_1,l_\infty}(x), T^{-1}_{l_1,l_\infty}(y))$.

According to this lemma, location problems with l_1 distance in the plane are equivalent to location problems with l_∞ distance simply by the given rotation and stretching T_{l_1,l_∞}.

The following lemma about the l_p norms in two dimensions will be used in Section 3.1.

Lemma 1.3 *Let $p \in [1,\infty]$. Let $x,y \in \mathbb{R}^2$ be such that $|x| = |y|$ and the angle between them is 90^0, i.e., $x_1y_1 + x_2y_2 = 0$. Then $l_p(x) = l_p(y)$.*

Proof: Choose r and α such that

$$\begin{aligned} x_1 &= r\ cos(\alpha) \text{ and} \\ x_2 &= r\ sin(\alpha). \end{aligned}$$

Then y can be given by $y = (y_1, y_2)$ with

$$\begin{aligned} y_1 &= r\ cos(\alpha + 90^0) = -r\ sin(\alpha), \\ y_2 &= r\ sin(\alpha + 90^0) = r\ cos(\alpha), \end{aligned}$$

such that we get

$$l_p(x) = \sqrt[p]{|r\ cos(\alpha)|^p + |r\ sin(\alpha)|^p} = l_p(y).$$

QED

Another distance measure (which is neither a norm nor a metric as the distance between two points can be infinite) will be introduced in the next definition.

Definition 5 *Let $x,y \in \mathbb{R}^n$. Then we define the* **horizontal distance** *between x and y as*

$$d_{hor}(x,y) = \begin{cases} |y_1 - x_1| & \text{if } x_i = y_i, \text{ for all } i = 2,3\ldots,n \\ \infty & \text{otherwise} \end{cases}$$

and the **vertical distance** *between x and y as*

$$d_{ver}(x,y) = \begin{cases} |y_n - x_n| & \text{if } x_i = y_i, \text{ for all } i = 1,2\ldots,n-1 \\ \infty & \text{otherwise.} \end{cases}$$

Consequently, for $n = 2$ the distances d_{hor} and d_{ver} are given as follows. Let $x, y \in \mathbb{R}^2$.

1. The vertical distance d_{ver} in the plane is given as

$$d_{ver}(x,y) = \begin{cases} |y_2 - x_2| & \text{if } x_1 = y_1 \\ \infty & \text{if } x_1 \neq y_1. \end{cases}$$

2. For the horizontal distance d_{hor} in the plane we get

$$d_{hor}(x,y) = \begin{cases} |y_1 - x_1| & \text{if } x_2 = y_2 \\ \infty & \text{if } x_2 \neq y_2. \end{cases}$$

To measure the distance between two points this definition does not seem to make much sense; as in most cases it will be infinity. But measuring the distance between a point and a line usually leads to finite vertical and horizontal distances. In the following chapters it will turn out that d_{ver} can be used to prove a lot of results, even for arbitrary norms. A generalization of the horizontal and vertical distance is the t-distance which will be introduced in Section 3.2.
The next lemma shows an obvious transformation from horizontal to vertical distances in $\mathbb{R}^2$.

Lemma 1.4 *For the transformation $T_{d_{hor},d_{ver}}$ given by a 90^0 rotation about the origin. i.e.,*

$$T_{d_{hor},d_{ver}} = \begin{pmatrix} 0 & -1 \\ 1 & 0 \end{pmatrix}$$

the following relations hold:

1. $d_{ver}(x,y) - d_{hor}(T_{d_{hor},d_{ver}}(x), T_{d_{hor},d_{ver}}(y))$ *and*

2. $d_{hor}(x,y) = d_{ver}(T_{d_{hor},d_{ver}}(x), T_{d_{hor},d_{ver}}(y))$.

Finally, we define the distance between two sets A and B:

Definition 6 *Let d be a arbitrary distance measure, i.e., a nonnegative function $d : \mathbb{R}^n \times \mathbb{R}^n \to \mathbb{R}$. Then the* **distance between two sets** *$A, B \subseteq \mathbb{R}^n$ is defined as follows.*

$$d(A,B) = \inf_{x \in A, y \in B} d(x,y).$$

1.3 Problem Description and Notation

Since we are mainly interested in the location of lines and hyperplanes, we repeat some of the necessary definitions and give some geometrical properties concerning lines and hyperplanes before we will formally describe the corresponding location problems.

Definition 7 *Let* $\mathfrak{n} = (s_1, s_2, \ldots, s_n) \in \mathbb{R}^n \setminus \{0\}$ *and* $b \in \mathbb{R}$ *be given. A* **hyperplane** $H = H_{\mathfrak{n},b} = H_{s_1,s_2,\ldots,s_n,b} \subseteq \mathbb{R}^n$ *is defined as*

$$\begin{aligned} H_{s_1,s_2,\ldots,s_n,b} &= \{x = (x_1, x_2, \ldots, x_n) : s_1x_1 + s_2x_2 + \cdots + s_nx_n + b = 0\} \\ &= \{x \in \mathbb{R}^n : \langle x, n \rangle + b = 0\}. \end{aligned}$$

The vector $\mathfrak{n} = (s_1, s_2, \ldots, s_n)$ *is orthogonal to* H. *It will be called a* **normal vector** *of* H.

The **normed normal vector** of the hyperplane H is given by

$$\mathfrak{n}^0 = \frac{1}{\sqrt{s_1^2 + \cdots + s_n^2}}(s_1, s_2, \ldots, s_n) = \frac{1}{\|n\|}(s_1, s_2, \ldots, s_n).$$

Note that $H_{s_1,\ldots,s_n,b} = H_{\alpha s_1,\ldots\alpha s_n,\alpha b}$ for any $\alpha \in \mathbb{R} \setminus \{0\}$, showing that the notation $H_{s_1,s_2,\ldots,s_n,b}$ is not unique for a given hyperplane H. But since the Euclidean length of $\mathfrak{n}^0$ is normed to 1 the notation $H_{\mathfrak{n}^0,b}$ is unique.

Obviously, hyperplanes in $\mathbb{R}^2$ are the usual **lines** $l \subseteq \mathbb{R}^2$ and we use the denotation

$$l_{s_1,s_2,b} = \{(x_1, x_2) : s_1x_1 + s_2x_2 + b = 0\}.$$

To get a unique description of the set of all lines in the plane, we could again require that $s_1^2+s_2^2 = 1$. But sometimes it will be more convenient to distinguish the following two cases.

Non-vertical lines : If the line $l_{s_1,s_2,b'}$ is not a vertical line, then $s_2 \neq 0$ and s_2 can therefore be normed to -1, such that we get

$$l_{s,-1,b} = \{(x_1, x_2) : x_2 = sx_1 + b\}$$

as the usual equation of a line with slope s and intercept b. When talking about vertical distances d_{ver} we only investigate non-vertical lines $l_{s,-1,b}$, $s, b \in \mathbb{R}$.

Vertical lines : If $s_2 = 0$ for a line $l_{s_1,s_2,b}$ then the line is parallel to the second unit vector e_2, i.e., after norming s_1 to -1 we get the equation for a vertical line through the point $(b, 0)$:

$$l_{-1,0,b} = \{(x_1, x_2) : x_1 = b\}.$$

Using only these two cases (i.e., we require that either $s_2 = -1$ or $s_2 = 0$ and $s_1 = -1$) the description of a line is unique. For any line $l = l_{s_1,s_2,b}$ we define

$$s = -\frac{s_1}{s_2} \in \mathbb{R} \cup \{\infty\}$$

as the **slope** of the line l which in the case of a vertical line can be infinite. The slope s_t of a two-dimensional vector $t = (t_1, t_2)$ is defined as the slope of the line passing through the origin $0 \in \mathbb{R}^2$ and containing t, i.e.,

$$s_t = \frac{t_2}{t_1} \in \mathbb{R} \cup \{\infty\}.$$

Definition 8 *For a hyperplane H define $\mathcal{B}_H^-$ and $\mathcal{B}_H^+$ as the* **two open half-spaces** *of* $\mathbb{R}^n$ **separated by the hyperplane** H *(without the hyperplane H itself, in each case).*

A hyperplane H **supports** a set A if $A \cap H \neq \emptyset$ and the whole set A lies on the same side of H, i.e., either $A \cap \mathcal{B}_{\mathcal{H}}^+ = \emptyset$ or $A \cap \mathcal{B}_{\mathcal{H}}^- = \emptyset$. A **tangent** to a set A is a line l which supports A. We also say that l **touches** A. Finally, a line l **separates** two points x and y if $x \in \mathcal{B}_l^-$ and $y \in \mathcal{B}_l^+$ or vice versa.

We are now ready to give a formal definition of the location problems discussed in this text. Given an index set $\mathcal{M} = \{1, 2, \ldots, M\}$ and for all $m \in \mathcal{M}$ an existing facility

$$Ex_m = (a_{m1}, a_{m2}, \ldots, a_{mn}) \in \mathbb{R}^n$$

with positive weight $w_m > 0$, find a new facility *New* such that

$$f(New) = \sum_{m \in \mathcal{M}} w_m d(Ex_m, New)$$

is minimized or such that

$$g(New) = \max_{m \in \mathcal{M}} w_m d(Ex_m, New)$$

is minimized, respectively. The function f is called the **sum** (or **median**) **objective**, whereas g is denoted **maximum** (or **center**) **objective function**.

Recall that in Definition 6 in the previous section the distance between a point $x \in \mathbb{R}^n$ and any set $New \subseteq \mathbb{R}^n$ has been defined as

$$d(x, New) = \inf_{y \in New} d(x, y).$$

We also define $W := \sum_{m \in \mathcal{M}} w_m$ as the sum of all weights.

If we are locating a line in the plane (i.e., *New* is a line l and $\mathbb{R}^n = \mathbb{R}^2$), then any line minimizing f is called a **median line**, and lines minimizing g are called **center lines**. The set of optimal lines is usually denoted by $\mathcal{L}^*$. Locating lines in the plane with respect to different distance functions is the subject of Chapters 2, 3, and 6. Instead of locating lines in the plane we can also look for optimal hyperplanes with respect to a given point set $\mathcal{E}x \subset \mathbb{R}^n$ in n-dimensional space. This will be done in Chapter 7. Interpreting *New* in the above formulas by other structures like circles, polygonal paths, trees, etc. leads to a variety of location problems, some of which will be discussed briefly in the last chapter. If *New* is a single point $x \in \mathbb{R}^2$, the above problem is the classical Weber or Fermat-Torricelli problem (for the sum objective function) or the classical center problem (for the maximum objective). Both problems are well-known in location theory, see e.g. the surveys of [Ham95, Pla95, LMW88].

In the following we will use the classification scheme of Hamacher and Nickel (see [HN96, HNS96]) for classical location problems which can easily be adapted to the location problems which we will discuss in the following chapters. The scheme consists of five positions $Pos1/Pos2/Pos3/Pos4/Pos5$ that are explained in the following. Tables 1.1–1.5 describe all symbols which are used in this text for the classification scheme.

Table 1.1. Position 1 of the classification scheme

Position 1	the new facility consists of ...
1	one point
$1l$	one line
rl	r lines
$1S$	one line segment
$1C$	one circle
$1curve$	one curve
$1poly$	one polygonal line
$1H$	one hyperplane

Table 1.2. Position 2 of the classification scheme

Position 2	decision space
$\mathbb{R}^2$	a planar problem
$\mathbb{R}^n$	a problem in n-dimensional space

Table 1.3. Position 3 of the classification scheme

Position 3	meaning of the symbol
$w_m = 1$	all weights are equal, i.e., $w_m = w$ for all $m \in \mathcal{M}$
R	the new facilities must not be located in the forbidden region R
R^D	a set of lines is restricted
$R = Polygon$	the forbidden region is a polygon
$\mathcal{E}x = \{L_1, \ldots, L_m\}$	the existing facilities are lines
$\cdot$	standard location problem with positive weights (without any additional constraint)

Table 1.4. Position 4 of the classification scheme

Position 4	the distance measure is ...
l_2	the Euclidean distance
l_1	the rectangular distance
l_∞	the Chebyshev distance
l_p	a p-norm
γ	a norm
γ_B	a block norm
$\tilde{\gamma}$	a gauge
$\tilde{\gamma}_B$	a polyhedral gauge
d_{hor}	the horizontal distance
d_{ver}	the vertical distance
d	any metric
γ_m	mixed norms (each of the existing facilities has its own norm)

Pos1 : This position gives the number and shape of new facilities. A number in this position without a specification of the shape always refers to point-shaped facilities.

Pos2 : The second entry specifies the type of the location problem with respect to the decision space. This information differentiates between continuous, network, and discrete location problems. Here we are only concerned with continuous location.

Pos3 : In this position particularities of the specific location problem are described. For example, any type of restriction can be defined here.

Pos4 : In continuous location this position is reserved for describing the distance measure between the existing and the new facilities (and also between the new facilities, if more than one facility is to be located).

Pos5 : The last position contains information on the objective function.

A dot in one of the positions indicates that no special assumptions are made. Note that we usually assume positive weights for the existing facilities (in contrast to [HN96, HNS96] who assume nonnegative weights $w_m \geq 0$ in the classification scheme). This is no restriction since, if some of the weights are zero, we simply can drop the corresponding facility and get an equivalent problem instance with positive weights. The case of equal weights $w_m = 1$ for all $m \in \mathcal{M}$ has to be specified in the third position of the classification scheme.
The following examples are given to demonstrate the usage of the classification scheme.

- First, the classical Weber (or Fermat-Torricelli) problem , i.e., to locate one point that minimizes the sum of Euclidean distances to a set $\mathcal{E}x$ of existing facilities is described by $1/\mathbb{R}^2/\cdot/l_2/\sum$.
- If in the classical Weber problem the distance measure d is used instead of the Euclidean distance, this is indicated by a d in Position 4 of the classification scheme, i.e., $1/\mathbb{R}^2/\cdot/d/\sum$.
- Analogously, $1/\mathbb{R}^2/\cdot/l_2/\max$ refers to the classical minmax or center problem with Euclidean distance.
- Locating a line in the plane with a distance measure which is derived from a norm γ is presented by $1l/\mathbb{R}^2/\cdot/\gamma/\sum$, if we want to minimize the median objective function, and by $1l/\mathbb{R}^2/\cdot/\gamma/\max$ in the case of the center objective.
- For the location of hyperplanes, we have to change $\mathbb{R}^2$ to $\mathbb{R}^n$ in the second position of the classification scheme and we have to refer to hyperplanes in the first position. Hence we get $1H/\mathbb{R}^n/\cdot/\gamma/\sum$ or $1H/\mathbb{R}^n/\cdot/\gamma/\max$, respectively.
- The expression $1l/\mathbb{R}^2/R^D/\gamma/\max$ refers to restricted line location problems with center objective function.
- Finally, if *New* is the new facility to be located in a planar median problem, this is indicated by $1New/\mathbb{R}^n/\cdot/\gamma/\sum$.
- A lot of other examples will be given in the following chapters.

Some further standard denotation which will be used in this text will be mentioned now. For a set $A \subseteq \mathbb{R}^n$ we use $conv(A), aff(A), int(A), \partial(A)$, $Ext(A)$, and $|A|$ as notation for the **convex hull**, the **affine hull**, the **interior**, the **boundary**, the **extreme points**, and the **number of elements** of A. If P is a convex polygon, $Ext(P)$ denotes the set of vertices of P. For a non-convex simple polygon P a vertex v is called **reflexive vertex** if $v \in int(conv(P))$. Finally, for a point $x \in \mathbb{R}^n$ and $\lambda \in \mathbb{R}$

$$\begin{aligned} A + x &= \{y + x : y \in A\} \text{ and} \\ \lambda A &= \{\lambda y : y \in A\} \end{aligned}$$

Table 1.5. Position 5 of the classification scheme

Position 5	the objective function is ...
$\sum$	the sum objective function
max	the center objective function
$\cdot$	refers to both objectives $\sum$ and max

describe the **translated** and the **scaled** set A. For a linear transformation $D : \mathbb{R}^n \to \mathbb{R}^n$ given as an $n \times n$ matrix, $det(D)$ denotes the **determinant** of D, which is nonzero if and only if D^{-1} exists.

The following properties are of special interest for line and hyperplane location problems. If they hold for a (continuous) line or hyperplane location problem, this problem can be reduced to a combinatorial optimization problem, and thus an optimal solution can be found in a finite candidate set . The properties will be discussed with respect to line location problems for some special distances, for general norms, gauges, and in the case of metrics in Chapters 2, 3, and 5. They will also appear in Chapter 7 for the n-dimensional case. A list of the abbreviations Med1, Med2, Cen1, and Cen2 for line location problems is given in Table 1.6 at the end of this section and also in the summary on page 181.

Med1 (Weak incidence property): There exists a median hyperplane passing through n affinely independent existing facilities.

Med2 (Pseudo-halving property): Every median hyperplane H^* satisfies

$$\sum_{Ex_m \in \mathcal{B}^-_{H^*}} w_m \leq \frac{W}{2}$$

$$\sum_{Ex_m \in \mathcal{B}^+_{H^*}} w_m \leq \frac{W}{2}.$$

Cen1 (Weak blockedness property): There exists a center hyperplane which is at maximum (weighted) distance from $n+1$ of the existing facilities.

Cen2 (Parallel facets property): If the weights are all equal, there exists a center hyperplane which is parallel to a facet of the convex hull of the existing facilities.

In Chapter 7 we will also use the following modified version of Cen1.

Cen1' (Modified weak blockedness property): There exists a center hyperplane which is at maximum (weighted) distance from $n+1$ affinely independent existing facilities.

Obviously, if Cen1' holds for a hyperplane location problem, then also Cen1 is true. Since Chapters 2 - 7 are dealing with the location of lines in $\mathbb{R}^2$ we list the four main properties again in the planar case.

Med1 (Weak incidence property): There exists a median line passing through two of the existing facilities.

Med2 (Pseudo-halving property): Every median line l^* satisfies

$$\sum_{Ex_m \in \mathcal{B}^-_{l^*}} w_m \leq \frac{W}{2}$$

$$\sum_{Ex_m \in \mathcal{B}_{l^*}^+} w_m \leq \frac{W}{2}.$$

Cen1 (Weak blockedness property): There exists a center line which is at maximum (weighted) distance from three of the existing facilities.

Cen2 (Parallel facets property): If the weights are all equal, there exists a center line which has the same slope as one of the facets of the convex hull of the existing facilities.

We used two different formulations of the first property Med1, because two points x, y with $x \neq y$ are always affinely independent. On the other hand, Cen1 and Cen1' really are two different properties, but we will show that they are equivalent for hyperplane location problems with distances derived from norms. In the summary on page 181 the differences between Cen1 and Cen1' are listed. Since in the literature the property Cen1 is more common than Cen1' and affine independency does not play an important role in the plane, we only discuss the property Cen1 when dealing with the location of lines. In Chapter 7 where we discuss hyperplane location problems, we use the stronger property Cen1'.

Some examples and remarks concerning the four properties will be added. Let us first discuss Med1 and Med2.

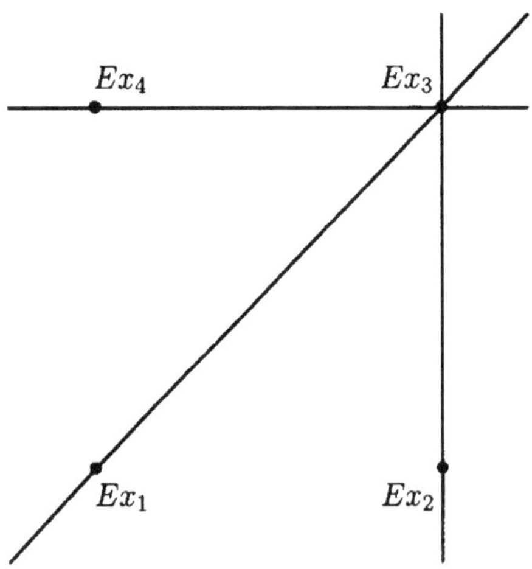

Figure 1.1. An example with four existing facilities and the set of candidate lines if Med1 and Med2 hold.

Example 1.1 *Consider the four vertices of a square $Ex_1 = (-1, -1)$, $Ex_2 = (1, -1)$, $Ex_3 = (1, 1)$, and $Ex_4 = (-1, 1)$ as our set $\mathcal{E}x$ of four existing facilities, graphed in Figure 1.1. Suppose that $w_1 = w_2 = 1$, $w_3 = 6$ and $w_4 = \frac{3}{2}$. If Med1* *holds for this line location problem, we only have to investigate the six lines passing through two of the existing facilities to find an optimal solution of the median problem. Furthermore, only three of these six lines satisfy the condition of Med2. Thus, if also Med2 holds for this problem, it is enough to consider the three lines graphed in Figure 1.1. For example, Med1 and Med2 are both true, if d is specified as the rectangular or the Euclidean distance (see Lemma 3.2 and Lemma 3.4 in Section 3.1). In the case of the rectangular distance, the line passing through Ex_3 and Ex_4 is the unique optimal solution. For the Euclidean distance, however, the line passing through Ex_1 and Ex_3 minimizes the median objective. This means that the optimal solutions of* $1l/\mathbb{R}^2/\cdot/l_1/\sum$ *and* $1l/\mathbb{R}^2/\cdot/l_2/\sum$ *do not coincide in this example.*

To formalize the conditions of Med2, we introduce Definition 10. Before doing this, we mention the following definition which has been introduced in [KM93] for Euclidean hyperplane location problems.

Definition 9 ([KM93]) *A hyperplane H which is the affine hull of n of the existing facilities and which satisfies both inequalities* $\sum_{Ex_m \in \mathcal{B}_H^-} w_m < \frac{1}{2}W$ *and* $\sum_{Ex_m \in \mathcal{B}_H^+} w_m < \frac{1}{2}W$ *is called a* **halving** *hyperplane.*

This definition has been motivated by the Euclidean norm: For l_2 [KM93] showed that all optimal hyperplanes are halving ones. For locating lines in the plane, in [MN80] the equivalent criterion

$$\left| \sum_{Ex_m \in \mathcal{B}_l^-} w_m - \sum_{Ex_m \in \mathcal{B}_l^+} w_m \right| < \sum_{Ex_m \in l} w_m$$

is used instead of $\sum_{Ex_m \in \mathcal{B}_l^-} w_m < \frac{1}{2}W$ and $\sum_{Ex_m \in \mathcal{B}_l^+} w_m < \frac{1}{2}W$. Also in [MN80] it has been shown to hold for all median lines in the Euclidean case. For distances derived from other norms this is not true in general. Therefore we need the following definition.

Definition 10 *A hyperplane H is called* **pseudo-halving**, *if*

$$\sum_{Ex_m \in \mathcal{B}_H^-} w_m \le \frac{W}{2} \quad \text{and}$$

$$\sum_{Ex_m \in \mathcal{B}_H^+} w_m \le \frac{W}{2}.$$

(Med2) can now be restated as: All optimal hyperplanes are pseudo-halving. Note that in Definition 10 it is not required that H passes through n of the existing facilities. Clearly, a halving hyperplane is also pseudo-halving and is the affine hull of n of the existing facilities. On the other hand, there exist hyperplanes H which are the affine hull of n of the existing facilities and

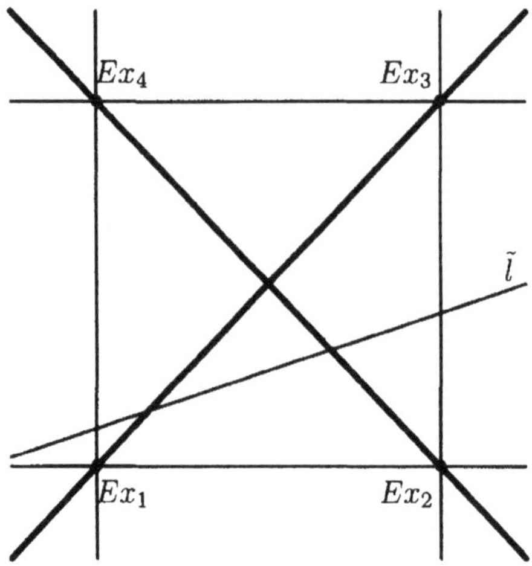

Figure 1.2. Difference between halving and pseudo-halving lines.

are pseudo-halving, and even optimal (!), but which are not halving, as the following example shows.

Example 1.2 *Consider the same set of existing facilities as in Example 1.1, but with equal weights* $w_m = 1, m = 1,2,3,4$. *For* $i < j$ *let* l_{ij} *denote the line passing through the existing facilities* Ex_i *and* Ex_j. *Furthermore, consider the line* $\tilde{l}$ *graphed in Figure 1.2. In the case that* $d = l_1$ *is the rectangular distance, all six lines* l_{ij} *lead to* $f(l_{ij}) = 4$ *and since Med1 holds for* l_1 *we conclude that these six lines are all optimal. Furthermore, also* $f(\tilde{l}) = 4$ *and thus* $\tilde{l}$ *is a median line. Since Med2 holds, all optimal lines are pseudo-halving, and thus* $\tilde{l}$ *and all six lines* l_{ij} *are pseudo-halving lines.*
Now we check if these lines are also halving. First consider $l \in \{l_{12}, l_{23}, l_{34}, l_{14}\}$. *Such a line* l *does not satisfy both*

$$\sum_{Ex_m \in \mathcal{B}_l^-} w_m < 2 \quad \text{and} \quad \sum_{Ex_m \in \mathcal{B}_l^+} w_m < 2$$

and therefore is not halving. For $\tilde{l}$ *it is true that*

$$\sum_{Ex_m \in \mathcal{B}_{\tilde{l}}^-} w_m < 2 \quad \text{and} \quad \sum_{Ex_m \in \mathcal{B}_{\tilde{l}}^+} w_m < 2,$$

but since $\tilde{l}$ *is not passing through two of the existing facilities, it also is not a halving line. This means that only the lines* l_{13} *and* l_{24} *are halving lines according to Definition 9. Hence there may exist optimal lines which are not halving.*

(Med1) is also called the weak incidence property in contrast to the **strong incidence property** that *all* optimal hyperplanes are the affine hull of n of the existing facilities. The strong incidence property will be discussed in Section 4.1 for lines and in Section 7.4 for hyperplane location problems.

For the center objective function, the weak blockedness property (Cen1) also is an existence statement. In Sections 4.2 and 7.4. we will investigate the **strong blockedness property** saying that *all* optimal hyperplanes are at maximum (weighted) distance from $n+1$ of the existing facilities.
Note that, if we write "at maximum distance" in the formulation of Cen1, we always refer to the weighted maximum distance $g(l) = \max_{m \in \mathcal{M}} w_m d(Ex_m, l)$.

Finally we discuss the last of the four properties, Cen2, in the plane. To this end we cite an additional definition from computational geometry, before we show an example for the parallel facets property.

Definition 11 (e.g. [Tou83]) *Let P be a convex polygon in the plane. A pair (e, v) consisting of a facet e and of a vertex v of P is called* **antipodal** *if there exist two different parallel supporting lines l^1, l^2 with $e \subseteq l^1$ and $v \in l^2$.*

This definition can also be extended to higher dimensions, see e.g. [KM93]. Antipodal pairs of $conv(\mathcal{E}x)$ will be used in Algorithm 3 to find center lines for problem instances with property Cen2. In the plane it is easy to verify the following connection between (Cen1) and (Cen2). If l is a center line which is parallel to a facet of the convex hull of the existing facilities then l is at maximum distance from three of the existing facilities, namely to the antipodal pair (e, v) where $e = (v_1, v_2)$ is a facet parallel to the line l and a translate of l is tangent to the vertex v. This means, that a problem instance (with equal weights $w_m = 1$, i.e., $1l/\mathbb{R}^2/w_m = 1/\cdot/\max$) with property (Cen2) also has property (Cen1), but not vice versa.

Example 1.3 *Consider Figure 1.3 where three existing facilities and their convex hull is graphed. If Cen2 holds, only the three marked candidate lines have to be evaluated to find an optimal solution of the center line location problem. Note that all three candidate lines are at maximum distance from Ex_1, Ex_2, and Ex_3. The three antipodal pairs are given by (e_m, Ex_m) for $m = 1, 2, 3$.*

1.4 Related Problems in Computational Geometry

In this section we discuss two problems from computational geometry, which are related to line and hyperplane location problems with center objective function. We also introduce some notation from computational geometry and prove some results which will be needed later.

1.4.1 *Transversal Theory*

The problem of finding a straight line minimizing the maximum (weighted) distance to a given point set is closely related to finding line transversals in

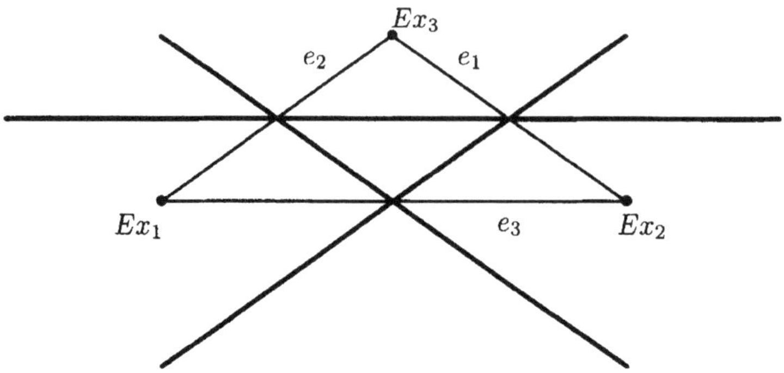

Figure 1.3. An example with three existing facilities and the set of candidate lines, if Cen2 holds.

Table 1.6. Possible properties of planar line location problems

Med1	There exists a median line passing through two of the existing facilities.
Med2	Every median line is pseudo-halving.
Cen1	There exists a center line at maximum (weighted) distance from three of the existing facilities.
Cen2	If the weights are all equal, there exists a center line parallel to a facet of the convex hull of the existing facilities.

computational geometry. We first define this problem and then discuss its relation to $1l/\mathbb{R}^2/\cdot/d/\max$.

Definition 12 (see e.g. [GPW93]) *Let $\mathcal{A} = \{A_1, A_2, \ldots, A_M\}$ be a family of convex sets in $\mathbb{R}^n$. Then l is called a* **line transversal** *of the family $\mathcal{A}$ if*

$$l \cap A_m \neq \emptyset \text{ for all } m \in \mathcal{M}.$$

A **k-transversal** *is an affine subspace of dimension k (a k-flat) which meets all the sets of $\mathcal{A}$.*

Instead of line transversal the denotation **stabbing line** is often used in computational geometry.

Given a family $\mathcal{A}$ of convex sets in $\mathbb{R}^n$, the **k-transversal problem** is to find out if there exist k-transversals for $\mathcal{A}$, and then to find all or some of these k-transversals. (For the special case that $k = 1$ the problem is called **line transversal problem** or **line stabbing problem** . This problem has been studied extensively in computational geometry, see e.g. [GPW93], or [Wen97] for a survey. In transversal theory it is usually assumed that all sets $A \in \mathcal{A}$ are convex. For $n = 2$ this is not a restriction, since in the plane all sets can be replaced by their convex hulls without changing the set of stabbing lines. Since we need this fact in Chapters 5 and 6 for line location, we will formulate it as the following lemma.

Lemma 1.5 *Let l be a line and A be a connected set. Then*

$$l \cap A = \emptyset \Longleftrightarrow l \cap conv(A) = \emptyset$$

Furthermore, l is a tangent to A if and only if l is a tangent to $conv(A)$.

Proof: Since $conv(A)$ is the smallest convex set containing A we know that

$$A \cap \mathcal{B}_l^- = \emptyset \quad \Longrightarrow \quad conv(A) \cap \mathcal{B}_l^- = \emptyset \tag{1.5}$$
$$A \cap (\mathcal{B}_l^- \cup l) = \emptyset \quad \Longrightarrow \quad conv(A) \cap (\mathcal{B}_l^- \cup l) = \emptyset \tag{1.6}$$

$\Longrightarrow$: Since $A \subseteq conv(A)$, $l \cap conv(A) = \emptyset$ implies $l \cap A = \emptyset$. For the second part assume that l is a tangent to A and that e.g. $A \cap \mathcal{B}_l^- = \emptyset$. Because of 1.5 we know that also $conv(A) \cap \mathcal{B}_l^- = \emptyset$, i.e., l is a tangent to $conv(A)$.

$\Longleftarrow$: Let $l \cap A = \emptyset$. Since A is connected this means that e.g. $A \cap (\mathcal{B}_l^- \cup l) = \emptyset$. From 1.6 it follows that $conv(A) \cap (\mathcal{B}_l^- \cup l) = \emptyset$ and hence l does not meet $conv(A)$. Finally, if l is a tangent to $conv(A)$ such that e.g. $conv(A) \cap \mathcal{B}_l^+ = \emptyset$, then from $A \subseteq conv(A)$ we conclude $A \cap \mathcal{B}_l^+ = \emptyset$. I.e., l is a tangent to A.

QED

[KM93] and [HII+93] already mention that H is a hyperplane transversal of the set

$$\mathcal{A} = \{\{x : w_m l_2(Ex_m, x) \leq 1\} : m \in \mathcal{M}\}$$

if the objective value of $1H/\mathbb{R}^n/\cdot/l_2/\max$ is equal to 1.
This statement will be made precise (and proven) in the following. To this end define for any real number $g \geq 0$ and any distance measure d

$$A_m(g) = \{x : w_m d(Ex_m, x) \leq g\} \text{ for all } m \in \mathcal{M}$$

as the set of all points whose weighted distance to Ex_m is less than or equal to g. Furthermore, let the family $\mathcal{A}(g)$ be given by

$$\mathcal{A}(g) = \{A_m(g) : m \in \mathcal{M}\}.$$

Recall that for a hyperplane H

$$g(H) = \max_{m \in \mathcal{M}} w_m d(Ex_m, H)$$

is the objective value of the center problem. Then we can formulate the following lemma.

Lemma 1.6 *Let $\mathcal{E}x = \{Ex_1, Ex_2, \ldots, Ex_M\}$ be a set of existing facilities with positive weights w_m and d be a distance measure such that all sets in $\mathcal{A}(g^0)$ are connected for $g^0 \geq 0$. Then the following holds for any hyperplane H:*

$$g(H) \leq g^0 \iff H \text{ is a hyperplane transversal for } \mathcal{A}(g^0).$$

Proof:

$$\begin{aligned} g(l) \leq g^0 &\iff w_m d(Ex_m, H) \leq g^0 \text{ for all } m \in \mathcal{M} \\ &\iff \text{for all } m \in \mathcal{M} \text{ there exists } x_m \in H : w_m d(Ex_m, x_m) \leq g^0 \\ &\iff H \cap A_m(g) \neq \emptyset \text{ for all } m \in \mathcal{M} \\ &\iff H \text{ is a hyperplane transversal for } \mathcal{A}(g). \end{aligned}$$

QED

As a conclusion we get the following theorem.

Theorem 1.2 *Let a set of existing facilities $\mathcal{E}x = \{Ex_1, Ex_2, \ldots, Ex_M\}$ with positive weights w_m be given. Then g^* is the objective value of $1H/\mathbb{R}^n/\cdot/d/\max$ if and only if g^* is the smallest real number g for which $\mathcal{A}(g)$ has a hyperplane transversal.*
In particular, a hyperplane H is a center hyperplane with objective value g^* if and only if H is a hyperplane transversal for $\mathcal{A}(g^*)$ and for all $g < g^*$ no hyperplane transversal for $\mathcal{A}(g)$ exists.

Proof: This follows directly from Lemma 1.6.

QED

We remark that the equivalence shown in Theorem 1.2 is similar to the equivalence of point location problems with center objective and *piercing problems* . For a definition of piercing problems, a proof of the equivalence result, and various solution approaches for $1l/\mathbb{R}^2/\cdot/\tilde{\gamma}/\max$ see [SW96].
The results of Lemma 1.6 and Theorem 1.2 will be needed in Section 4.2 and in the whole of Chapter 5. More precisely, we will use these results to

- transform algorithms from line/hyperplane location to transversal problems as will be done for the fast block norm algorithm in Section 7.3

  and, on the other hand we also need it

- to establish algorithms for some center line location problems, especially for distance functions which are not derived from a norm, see Section 5.4.

1.4.2 Set Width Problems

The **set width problem** can be described as follows: Given a set $\mathcal{E}x$ of (unweighted) points in $\mathbb{R}^n$, find the minimal Euclidean distance between two parallel hyperplanes H^1, H^2 such that $\mathcal{E}x \subseteq \mathcal{B}^-_{H^1} \cap \mathcal{B}^+_{H^2}$, i.e., all existing facilities lie between the two parallel hyperplanes H^1 and H^2.

The set width problem also is equivalent to $1H/\mathbb{R}^n/w_m = 1/l_2/\max$. This has first been proven in [KM93, HII+93] and will formally be stated in the next lemma.

Lemma 1.7 ([KM93, HII+93]) *Given a set $\mathcal{E}x$ of points in $\mathbb{R}^n$, the optimal value of the set width problem is equal to g^0 if and only if the optimal value of $1H/\mathbb{R}^n/w_m = 1/l_2/\max$ is equal to $\frac{g^0}{2}$.*

Now we define a modified version of the set width problem, which we call the **set width problem with norm γ** as follows.
Given a norm γ with distance measure d and a set of points $\mathcal{E}x$, find the minimal distance $d(H^1, H^2)$ between two parallel hyperplanes H^1, H^2 such that $\mathcal{E}x \subseteq \mathcal{B}^-_{H^1} \cap \mathcal{B}^+_{H^2}$. (Recall that the distance between two sets has been given in Definition 6).

Then, using the methods which we will develop in Sections 3.2 and 7.2 it is easy to show that the set width problem with norm γ and the hyperplane location problem $1H/\mathbb{R}^n/w_m = 1/\gamma/\max$ are equivalent, too. Since we do not focus on set width problems we only mention this result here.

Lemma 1.8 *Given a set $\mathcal{E}x$ of points in $\mathbb{R}^n$, the optimal value of the set width problem with norm γ is equal to g^0 if and only if the optimal value of $1H/\mathbb{R}^n/w_m = 1/\gamma/\max$ is equal to $\frac{g^0}{2}$.*

1.5 Piecewise Linear Programs with Restrictions

First we will define the median of a set of real numbers. This definition is well known mainly in statistics.

Definition 13 (e.g. [Hay81, Kre91]) *Let $c_1, c_2, \ldots, c_M$ be a set of real numbers and let $w_1, w_2, \ldots, w_M$ be some positive weights, $w_m \geq 0, m = 1, \ldots, M$. Furthermore define $W = \sum_{m \in \mathcal{M}} w_m$ as the sum of all weights. Then*

$$median\{(c_m, w_m) : m \in \mathcal{M}\} = \left\{ x : \sum_{m:c_m < x} w_m \leq \frac{W}{2} \text{ and } \sum_{m:c_m > x} w_m \leq \frac{W}{2} \right\}$$

is the **set of medians** *for the weighted set of numbers c_m.*

Note that in most cases the median is unique, but it is an interval if there exists c_k such that $\sum_{m:c_m \leq c_k} w_m = \frac{W}{2}$. We also remark that there always exists a $k \in \mathcal{M}$ such that $c_k \in median\{c_m, w_m\}$.

Now we start with the simple case of one-dimensional piecewise linear programs. These provide an important tool for solving classical median facility problems $1/\mathbb{R}^2/\cdot/l_1/\sum$ with rectangular distance, since this location problem can be separated into two independent subproblems which are both piecewise linear and convex optimization problems in one variable (see [Ham95]). In Chapters 2, 5, 7, and in Chapter 8 we will use the following properties of one-dimensional convex and piecewise linear functions. Many versions of Lemma 1.9 have already been used in optimization literature, often in an implicit way.

Lemma 1.9 *Let $\mathcal{M} = \{1, 2, \ldots, M\}$ and*

$$f(x) = \sum_{m \in \mathcal{M}} w_m |c_m - x|$$

with weights $w_m \geq 0$. Then the following statements hold.

1. *[Hay81, Ham95] $f(x)$ is minimal if and only if $x \in median\{(c_m, w_m) : m \in \mathcal{M}\}$. In particular, there exists a $k \in \mathcal{M}$ such that c_k minimizes f.*
2. *If $x \neq c_m$ for all $m \in \mathcal{M}$ then there exists a $k \in \mathcal{M}$ such that $f(c_k) \leq f(x)$ and there is no other point $c_m, m \in \mathcal{M}$ between c_k and x.*

Proof: Note that f is a piecewise linear, convex function with breakpoints $c_m, m \in \mathcal{M}$. Now suppose x is optimal, but not a median of $\{(c_m, w_m) : m \in \mathcal{M}\}$. Without loss of generality assume that

$$\sum_{m: c_m < x} w_m > \frac{W}{2}.$$

Then there exists a small $\epsilon > 0$ such that

$$\begin{aligned} \{m : c_m < x\} &= \{m : c_m < x - \epsilon\} \text{ and} \\ \{m : c_m \geq x\} &= \{m : c_m \geq x - \epsilon\}. \end{aligned}$$

We calculate

$$\begin{aligned} f(x-\epsilon) &= \sum_{m: c_m < x} w_m(x - \epsilon - c_m) + \sum_{m: c_m \geq x} w_m(c_m - x + \epsilon) \\ &= \sum_{m: c_m < x} w_m(x - c_m) + \sum_{m: c_m \geq x} w_m(c_m - x) \\ &\quad + \epsilon \left(\sum_{m: c_m \geq x} w_m - \sum_{m: c_m < x} w_m \right) \\ &< f(x) + \epsilon \left(\frac{W}{2} - \frac{W}{2} \right) = f(x) \end{aligned}$$

and see that x cannot be optimal. For $x \notin median\{(c_m, w_m) : m \in \mathcal{M}\}$ the above equation also proves the second statement.

If, on the other hand, x is a median of the weighted set $\{(c_m, w_m)\}$ we similarly calculate for all sufficiently small $\epsilon > 0$ that

$$f(x-\epsilon) \geq f(x) \text{ and } f(x+\epsilon) \geq f(x).$$

Since f is a convex function, this means that x is optimal. Furthermore, if x is no breakpoint of f but a median, we finally get

$$f(x-\epsilon) = f(x) = f(x+\epsilon),$$

completing the proof.

QED

Now let us pay attention to the n-dimensional case, $n \geq 2$. For $n = 2$ the resulting problems also play an important role in classical facility location. For example, all location problems with polyhedral gauges, i.e., problems of the type $1/\mathbb{R}^2/\cdot/\tilde{\gamma}_B/\sum$ and of the type $1/\mathbb{R}^2/\cdot/\tilde{\gamma}_B/\max$ can be written as piecewise linear programs in two dimensions and therefore be solved efficiently, see [Ham95, Nic95]. In the context of line location problems we will show in Chapter 2 that line location problems with the vertical distance d_{ver} are also convex and piecewise linear optimization problems for the median and for the center objective function.

In the following we will first show some general properties of piecewise linear convex problems and then use this theory to solve such problems also in the case that (non-convex) constraints are added. These results will be needed in Chapter 6 for the discussion of restricted line location problems. Based on the results of this section, we will develop efficient solution approaches for line location problems with forbidden polyhedral areas, i.e., for $1l/\mathbb{R}^2/R = Polygon/\gamma/\cdot$, see Algorithms 8 and 9 in Section 6.3.

Some of the concepts of the remainder of this section, especially the concept of bumpy sets in the area of piecewise linear programs, are based on previous results obtained by the author together with Nickel, see [NS97]. In contrast to that work, most proofs in this section are done in a more geometrical way. We consider the following piecewise linear optimization problem in $\mathbb{R}^n$:

$$\text{(OL)} \quad \min_{x \in \mathbb{R}^n} f(x) \ ,$$

where $f(x) := \max\{f_1(x), f_2(x), \ldots, f_K(x)\}$ and the $f_i, i = 1, 2, \ldots, K$ are K different affine functions. This is a well-known convex problem that can be solved efficiently by linear programming methods. Without loss of generality we assume that none of the functions f_i, $i = 1, \ldots, K$ is redundant, i.e., for all $i = 1, \ldots, K$ there exists some $x \in \mathbb{R}^n$ such that

$$f_i(x) > \max_{k=1,2,\ldots,K, k \neq i} f_k(x).$$

(Note that if this is not the case, such a redundant function can be deleted.)

Then the **cells** $C_i, i = 1, 2, \ldots, K$ defined by

$$C_i = \{x \in \mathbb{R}^n : f_i(x) \geq f_k(x), \forall k \in \{1, \ldots, K\}\},$$

are non-empty. The corresponding **cell partition** , i.e., the set of all cells is denoted by $\mathcal{C}$. Some straightforward properties of $\mathcal{C}$ are summarized below.

1. All $C \in \mathcal{C}$ are n-dimensional polyhedral sets.
2. $\bigcup_{k=1,\ldots,K} C_k = \mathbb{R}^n$, and
3. $int(C_i) \cap int(C_j) = \emptyset \ \ \forall C_i \neq C_j, C_i, C_j \in \mathcal{C}$.

We remark that, even if the functions f_k, $k \in \{1, \ldots, K\}$, are not explicitely given, but we have a cell partition $\mathcal{C}$ of $\mathbb{R}^n$ and a convex function f which is linear on each cell $C \in \mathcal{C}$, we can apply the theory which we will develop in the following. To denote the set of optimal solutions of (OL) we introduce

$$\begin{aligned} t^* &:= \min_{x \in \mathbb{R}^n} f(x) \text{ and} \\ \mathcal{X}^* &:= \{x \in \mathbb{R}^n : f(x) = t^*\}. \end{aligned}$$

The concept of cells has been frequently used in optimization (see [RS72b, Ede87, HN95, Nic95] and references therein). From this literature the following result is well known.

Theorem 1.3 *$\mathcal{X}^*$ is either a whole cell or an r-dimensional face of a cell, $r \in \{0, 1, \ldots, n-1\}$. Especially, there exists a point $x \in \mathcal{X}^*$ which is an extreme point of a cell $C \in \mathcal{C}$.*

Proof: We can solve (OL) by examining all cells. For each cell $C \in \mathcal{C}$ the optimization problem is a linear program with the polyhedral feasible set C. Therefore the result follows from well known properties of linear programming theory, see e.g. [NW88].

QED

We define $\mathcal{C}^\partial$ as the set of all facets of all cells $C \in \mathcal{C}$.
It will sometimes be convenient to use a set of simpler structures, e.g. linear subspaces instead of the facets of all cells $\mathcal{C}^\partial$. Such sets are introduced in the next definition.

Definition 14 *Consider a cell partition $\mathcal{C}$ of $\mathbb{R}^n$ and let $\mathcal{C}^\partial$ be the set of all facets of the cells in $\mathcal{C}$. Then a set of hyperplanes $\mathcal{H}$ is called a* **set of construction hyperplanes** *for $\mathcal{C}$, if for each $h \in \mathcal{C}^\partial$ there exists an $H \in \mathcal{H}$ such that $h \subseteq H$.*

For this section we define $\mathcal{H}$ as the set of all hyperplanes H_{ij} with

$$H_{ij} = \{x : f_i(x) = f_j(x)\}$$

for all $i, j \in \{1, 2, \ldots, K\}$ with $i \neq j$. $\mathcal{H}$ is a set of construction hyperplanes according to Definition 14. In the n–dimensional case the construction hyperplanes are $(n-1)$–dimensional sets; in the plane we also denote them by **construction lines**. Looking at the set of points on construction hyperplanes

$$\mathcal{P}(\mathcal{H}) := \{x : \text{ there exists } H \in \mathcal{H} \text{ such that } x \in H\},$$

we note that the set of cell facets $\mathcal{C}^\partial$ leads to the smallest possible set $\mathcal{P}$, i.e.,

$$\mathcal{P}(\mathcal{C}^\partial) \subseteq \mathcal{P}(\mathcal{H})$$

for all sets of construction hyperplanes $\mathcal{H}$. This means, $\mathcal{C}^\partial$ is the smallest possible set of construction hyperplanes in this sense.

One more definition will be needed for the following considerations. For $t \in \mathbb{R}$ define the **level set** $L_\leq(t)$ and the **level curve** $L_=(t)$ as

$$\begin{aligned} L_\leq(t) &= \{x \in \mathbb{R}^n : f(x) \leq t\}, \\ L_=(t) &= \{x \in \mathbb{R}^n : f(x) = t\}. \end{aligned}$$

Note that for any convex function f, $L_\leq(t)$ is a closed, convex set which is nonempty for all $t \geq t^*$.

Lemma 1.10 *The level curves $L_=(t)$ are piecewise linear on $\mathcal{C}$, i.e., in every cell $C \in \mathcal{C}$ the level curve $L_=(t) \cap C$ has ones of the following shapes:*

1. *empty*
2. *the whole cell C (then $t = t^*$),*
3. *an r–dimensional face of $C, r \in \{0, 1, \ldots, n-1\}$,*
4. *the intersection of C and a hyperplane H with $int(C) \cap H \neq \emptyset$ (then it follows that $t > t^*$).*

Proof: Since the functions $f_i, i = 1, 2, \ldots, K$, are linear, their non-empty level curves are hyperplanes or the whole $\mathbb{R}^n$. This means that $L_=(t) \cap C$ is either empty or the intersection of C and a hyperplane or the whole cell C. The latter only can occur if $f(x) = t^*$ for all $x \in C$, by the convexity of f. Furthermore, if an interior point of a cell is included in $L_=(t)$, but not the whole cell is, then t cannot be the optimal value.

QED

Lemma 1.11 *For any $t \geq t^*$ and any set $\mathcal{H}$ of construction hyperplanes*

$$Ext(L_\leq(t)) \subseteq \mathcal{P}(\mathcal{H}).$$

Proof: Let $x \in Ext(L_\leq(t))$ be an extreme point of a level set with $x \in C$ for a cell $C \in \mathcal{C}$. Since the level curves $L_=(t)$ are linear on C (see Lemma 1.10), no extreme point of $L_{\leq}(t)$ can be in the interior of a cell, i.e., $x \notin int(C)$. Consequently, there exists a $c \in \mathcal{C}^{\partial}$ with $x \in c$, showing that $x \in \mathcal{P}(\mathcal{C}^{\partial})$.

QED

Now we use the theory developed above to discuss restricted piecewise linear minimization problems. To this end, let $R \subseteq \mathbb{R}^n$ be a **restricting set**. Then we add the constraint $x \notin int(R)$ and get (ROL) from (OL) as

$$\text{(ROL)} \quad \min_{x \notin int(R)} f(x).$$

With this additional constraint, (ROL) is not necessarily convex. Nevertheless, with the help of geometric properties we will be able to derive structural results for this problem for a large class of different sets R. As mentioned before, these results will be needed in Chapter 6 for the discussion of restricted line location problems.

For (ROL) we introduce the following notation.

$$\begin{aligned} t_R^* &:= \min_{x \in \mathbb{R}^n \setminus int(R)} f(x) \text{ and} \\ \mathcal{X}_R^* &:= \{x \in \mathbb{R}^n : x \notin int(R) \text{ and } f(x) = t_R^*\}. \end{aligned}$$

Using level sets and level curves, we can reformulate (OL) and (ROL) (as done in [Ham95] for planar location problems).

Lemma 1.12 *(OL) and (ROL) can be formulated as follows.*

1. $t^* = \min\{t : L_=(t) \neq \emptyset\}$ *if and only if* t^* *is the optimal objective value of (OL).*
2. $t_R^* = \min\{t : L_=(t) \cap (\mathbb{R}^n \setminus int(R)) \neq \emptyset\}$ *if and only if* t_R^* *is the optimal objective value of (ROL).*

The following theorem is a consequence of the convexity of f (see also [HT96, HN94] and [Ham95] in the context of location theory).

Theorem 1.4 *If* $\mathcal{X}^* \subseteq int(R)$ *then* $\mathcal{X}_R^* \subseteq \partial R$.

Proof: Let $x \in \mathcal{X}^*$ and assume that there exists a $y \in \mathbb{R}^n$ such that $y \in \mathcal{X}_R^* \backslash \partial R$. Then $y \notin R$, yielding that $f(x) < f(y)$. Therefore we get for all $\lambda \in (0,1]$

$$\begin{aligned} f(\lambda x + (1-\lambda)y) &\leq \lambda f(x) + (1-\lambda)f(y) \\ &< \lambda f(y) + (1-\lambda)f(y) \\ &= f(y). \end{aligned}$$

This also holds if we choose λ such that $\lambda x + (1-\lambda)y \in \partial R$, which proves the theorem.

QED

The main result for the problem (ROL) with a convex restricting set $R \subseteq \mathbb{R}^n$ can now be proved.

Theorem 1.5 *Let R be a convex set, $R \subseteq \mathbb{R}^n$ and $\mathcal{X}^* \subseteq int(R)$. Then there exists an $x_R^* \in \mathcal{X}_R^*$ and $c \in \mathcal{C}^\partial$ with*

$$x_R^* \in (c \cap \partial R) \ .$$

<u>Proof:</u> Let t_R^* be the optimal value of the restricted problem. From Lemma 1.12 we know that

$$L_\leq(t_R^*) \subseteq R \text{ and } L_=(t_R^*) \cap \partial R \neq \emptyset,$$

i.e., the polyhedral set $L_\leq(t_R^*)$ touches the convex set R from inside. This happens in at least one extreme point of the polyhedral set (see, e.g., [Sha78]), meaning that there exists an $x \in Ext(L_\leq(t_R^*)) \cap \partial R$. With Lemma 1.11 we conclude that

$$x \in \mathcal{P}(\mathcal{C}^\partial) \cap \partial R.$$

QED

In the next theorem we concentrate on $n = 2$. Furthermore, we replace the set of all cell facets $\mathcal{C}^\partial$ by a set of construction lines $\mathcal{H}$, since $\mathcal{H}$ is easier to compute.

Theorem 1.6 *Let $R \subseteq \mathbb{R}^2$ be a convex set such that $\mathcal{X}^* \subseteq int(R)$. Then there exists an optimal solution $x_R^* \in \mathcal{X}_R^*$ in the finite set of points*

$$Cand = \{H \cap \partial R : H \in \mathcal{H} \text{ and } dim(H \cap \partial R) = 0\}.$$

<u>Proof:</u> Let $x_R^* \in \mathcal{P}(\mathcal{C}^\partial) \cap \partial R$ be an optimal solution of the restricted problem according to Theorem 1.5. We have to show that there exists an $H \in \mathcal{H}$ such that $x_R^* \in H$ and $dim(H \cap \partial R) = 0$. To this end we consider

$$\mathcal{V} = \{Ext(L_\leq(t)) : t^* \leq t \leq t_R^*\}.$$

$\mathcal{V}$ has the following properties.

1. $\mathcal{V} \subseteq R$.

2. Since f is continuous, there exists $v \in \mathcal{V}$ and a path P from v to x_R^* such that $P \subseteq \mathcal{V}$.

3. $\mathcal{V} \subseteq \mathcal{P}(\mathcal{C}^\partial)$.

Thus $\mathcal{V}$ consists of linear pieces. Choose a linear piece s of $\mathcal{V}$, $s \subseteq P$ between x_R^* and another point $v' \in int(R)$. Then let $H \in \mathcal{H}$ be a construction line containing s (hence $x_R^* \in H$). Since R is convex and H is a line passing through an interior point of R we know from Theorem 3.4. in [Sha78] that H intersects ∂R in at most two points and therefore $dim(H \cap \partial R) = 0$.

QED

Using an induction argument, it is possible to improve the result of Theorem 1.5 (for n dimensions), and for polygonal sets in $\mathbb{R}^n$ further results can be obtained (see [NS97]). Since we will not refer to these results, we state only one sharper characterization without proof.

Theorem 1.7 ([NS97]) *Let $R \subseteq \mathbb{R}^n$ be convex, and $\mathcal{X}^* \subseteq int(R)$. Then there exists an optimal solution $x_R^* \in \mathcal{X}_R^*$ which is a zero-dimensional intersection between the boundary ∂R and at most $n-1$ construction lines $H \in \mathcal{H}$.*

Finally we consider special classes of non-convex forbidden sets which may have more practical applications. While reviewing the proof of Theorem 1.5 it becomes apparent that we get a finite candidate set for the optimal solution of (ROL) whenever we have only a finite set of points on ∂R which can lie in the (relative) interior of linear pieces of a level curve. Therefore we introduce the concept of bumpy sets, which was first mentioned in [Ham95] in location theory, see also [HN95]. The definition given here is a slightly modified version of the ones given in these references.

Definition 15 *A* **bumpy set** *$R \subseteq \mathbb{R}^2$ with its set of* **roots** *Roots(R) and its set of* **bumps** *Bumps(R) is any set which can be constructed by a finite application of the following rules:*

1. *Any convex set R is a bumpy set with $Roots(R) = \emptyset$ and $Bumps(R) = \emptyset$.*

2. *If R_0 is a bumpy set then $R = R_0 \cup R_1$ is a bumpy set if*

   - *$conv(R_1) \subseteq R$ and*
   - *there exist exactly two points $r_1, r_2 \in \partial R_0 \cap \partial R_1 \cap \partial R$.*

   In this case R_0 is called the **base** *of R and R_1 is called the new* **bump**. *The two points r_1, r_2 are the new roots of R, i.e.;*

$$\begin{aligned} Roots(R) &= Roots(R_0) \cup \{r_1, r_2\} \text{ and} \\ Bumps(R) &= Bumps(R_0) \cup \{R_1\}. \end{aligned}$$

Note that, depending on the construction of the bumpy set, the same bumpy set can have different sets of roots, i.e., $Roots(R)$ is not uniquely defined. (The same holds for $Bumps(R)$.) The following results, however, hold for any set of roots. For algorithmic purposes the smallest set of roots is preferable.

In [NS97] some examples and applications of bumpy sets can be found. One example is that all simple polygons are bumpy sets. Recall that a vertex v of a simple polygon P is called a *reflexive vertex* if $v \in int(conv(P))$. The set of roots of a simple polygon R is then given by

$$Roots(R) = \{v : v \text{ is a reflexive vertex of } R\}.$$

Lemmas 1.14 and 1.15 will characterize the set of roots of a bumpy set. They are needed to adapt the proof of Theorem 1.5 to the case of bumpy sets. To prove these two lemmas we need the following property of convex sets.

Lemma 1.13 *Let R be a convex set and s be a line segment. If $s \subseteq R$ and $int(s) \cap \partial R \neq \emptyset$, then $s \subseteq \partial R$.*

Proof: Let $int(s) \cap \partial R \neq \emptyset$. Consider the line l containing the line segment s. Since R is convex, l is a tangent to R (see [Sha78]) and we conclude that $l \subseteq \mathbb{R}^2 \setminus int(R)$. Together with $s \subseteq R$ we conclude that $s \subseteq \partial R$.

QED

Lemma 1.14 *Let R be a bumpy set and s be a line segment. If $s \subseteq R$ and $int(s) \cap \partial R = \{x\}$, then $x \in Roots(R)$.*

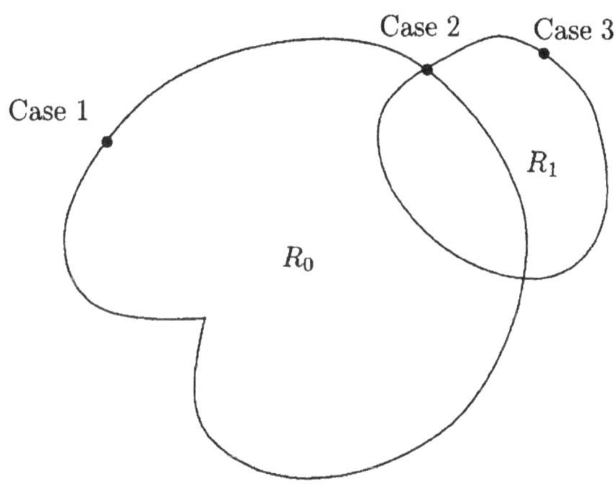

Figure 1.4. Three possible cases in the induction proof of Lemma 1.14.

Proof: We will use induction over the number of bumps.
For a convex set R and a line segment $s \subseteq R$ we know from Lemma 1.13 that if $int(s) \cap \partial R \neq \emptyset$ then $s \subseteq \partial R$, i.e., the interior of s cannot touch ∂R from inside at a unique point.
Taking any bumpy set R with base R_0 and bump R_1, let $s \subseteq R$ be a line segment with $int(s) \cap \partial R = \{x\}$. Now we distinguish three cases, using that

$$\partial R \subseteq \partial R_0 \cup \partial R_1,$$

(see Figure 1.4):

Case1 : If $x \notin \partial R_1$, we know that $x \in Roots(R_0) \subseteq Roots(R)$ due to the induction hypothesis.

Case 2 : If $x \in \partial R_1 \cap \partial R_0$, then by definition $x \in \partial R$ is one of the (new) roots of R.

Case 3 : If $x \in \partial R_1 \backslash \partial R_0$, then $x \in \partial conv(R_1)$ as $conv(R_1) \subseteq R$. Therefore we have a linear piece $s \subseteq conv(R_1)$ with $int(s) \cap \partial(R) = \{x\}$, a contradiction to Lemma 1.13.

QED

Lemma 1.15 *Let R be a bumpy set and $s \subseteq R$ a line segment. If $s' := s \cap \partial R$ is a line segment with $s \neq s'$, then there exists a root $x \in s'$.*

Proof: Again we use induction over the number of bumps.
For a convex set R and line segments s, s' satisfying the assumptions of Lemma 1.15 we again use Lemma 1.13 to conclude that $s = s'$ in this case.
Now take any bumpy set R with base R_0 and bump R_1. Let $s, s' \subseteq R$ be line segments with $s' = s \cap \partial R$ and $s' \subset s$. Again we distinguish three cases.

Case1 : If $s' \subseteq \partial R_0$, we use the induction hypothesis to conclude that the endpoints of s' are roots.

Case 2 : If $s' \cap \partial R_1 \neq \emptyset$ and $s' \cap \partial R_0 \neq \emptyset$, then there exists a point $x \in s'$ with $x \in \partial R_0 \cap \partial R_1 \cap \partial R$, i.e., $x \in Roots(R)$.

Case 3 : If $s' \subseteq \partial R_1 \setminus \partial R_0$, then $s' \subseteq \partial conv(R_1)$. Therefore we have $int(s) \cap \partial conv(R_1) \neq \emptyset$ and thus $s = s'$ according to Lemma 1.13.

QED

Finally, the following theorem shows that in order to solve (ROL) with a bumpy set R it suffices to examine the candidate set defined in Theorem 1.6, i.e., the zero-dimensional intersections between the boundary of R and the construction lines $H \in \mathcal{H}$ and, additionally, the roots of the bumpy set R.

Theorem 1.8 *Let $R \subseteq \mathbb{R}^2$ be a bumpy set such that $\mathcal{X}^* \subseteq int(R)$. Then there exists an optimal solution $x_R^* \in \mathcal{X}_R^*$ in the finite set of points*

$$Cand_{bumpy} = Cand \cup Roots(R).$$

Proof: Let t_R^* be the optimal value of the restricted problem. From Lemma 1.12 we know that

$$L_{\leq}(t_R^*) \subseteq R \text{ and } L_{=}(t_R^*) \cap \partial R \neq \emptyset.$$

Then there exists a facet s of $L_{\leq}(t_R^*)$ (i.e., s is a maximal linear piece of $L_{=}(t_R^*)$) such that $s \subseteq R$ and $s \cap \partial R \neq \emptyset$. The following cases are possible.

- Either there exists an endpoint x of the line segment s such that $x \in \partial R \cap s$. Then x is an extreme point of $L_{\leq}(t_R^*)$, and therefore $x \in Cand$ (see Lemma 1.11)

- Or there exists a segment s^0 contained in s such that $int(s^0) \cap \partial R = \{x\}$; then $x \in Roots(R)$ according to Lemma 1.14.

- Or there exists a segment $s' \neq s$ contained in s such that $s' \subset \partial R$; then Lemma 1.15 shows that the there exists a root in s'.

QED

If we apply Theorem 1.8 to simple, not necessarily convex polygons, we get immediately the following corollary.

Corollary 1.1 *Let R be a polygon and let $\mathcal{X}^* \subseteq int(R)$. Then there exists an optimal solution $x_R^* \in \mathcal{X}_R^*$ in the finite set of points*

$$Cand_{Polygon} = Cand \cup \{\textit{reflexive vertices of } R\}.$$

2 LINE LOCATION WITH VERTICAL DISTANCE

2.1 Results for the Vertical Distance

In this section we consider line location problems with vertical distance d_{ver}. The vertical distance between a point $x = (x_1, x_2)$ and a non-vertical line $l = l_{s,-1,b}$ with slope $s \in \mathbb{R}$ and intercept $b \in \mathbb{R}$ is given by the length of the vertical line segment between x and l and can be calculated by

$$\begin{aligned} d_{ver}(x, l) &= \min_{y \in l} d_{ver}(x, y) \\ &= d_{ver}(x, (x_1, y_2)) \text{ with } y = (x_1, y_2) \in l \\ &= |x_1 s - x_2 + b|. \end{aligned}$$

If l is a vertical line and $x \notin l$, then $d_{ver}(x, l) = \infty$, while $d_{ver}(x, l) = 0$ for $x \in l$.

For line location problems with $\mathcal{E}x = \{Ex_1, Ex_2, \ldots, Ex_M\}$ as the set of existing facilities it follows that a vertical line can never be optimal unless all existing facilities have the same first coordinate $a_{m1} = a_1$ for all $m \in \mathcal{M}$. But this case is trivial and will therefore be neglected. Note that then all properties Med1, Med2, Cen1, and Cen2 hold; Med1 and Cen1 even in the strong sense. Summarizing, the objective functions of the median and the center problem with vertical distance are given by

$$f(l_{s,-1,b}) = \sum_{m \in \mathcal{M}} w_m |a_{m1} s - a_{m2} + b| \text{ and}$$

$$g(l_{s,-1,b}) = \max_{m \in \mathcal{M}} w_m |a_{m1}s - a_{m2} + b|, \text{ respectively.}$$

Note that both functions are convex in the two variables s and b.

There are many applications showing the importance of line location problems with vertical distance. The median problem $1l/\mathbb{R}^2/\cdot/d_{ver}/\sum$ is known in statistics as *L_1-approximation* and the center problem $1l/P/\cdot/d_{ver}/\max$ is called *Chebyshev or L_∞-approximation in two variables.* Note that minimizing the squared sum of vertical distances to the existing facilities $(1l/\mathbb{R}^2/\cdot/(d_{ver})^2/\sum)$ is the same as the well-solved problem of finding a regression line (see e.g., [Hay81]).
The literature dealing with line location problems with vertical distance $1l/\mathbb{R}^2/\cdot/d_{ver}/\sum$ will now be briefly described. In the following a survey to the statistics literature will be given. In some older work, e.g., in [Kar58] and in [Sch73] infinite iterative processes are applied to find a median line minimizing the sum of weighted vertical distances (a so-called least weighted absolute deviations line), but not even convergence of the approach in [Sch73] is guaranteed. Another approach used in statistics is to formulate and solve $1l/\mathbb{R}^2/\cdot/d_{ver}/\sum$ as a linear programming problem (see [CCF55, Fis61, Wag59]). The advantage of this method is that it can also be applied in the higher dimensional case, but again, no polynomial bounds on the running time are given. In [MN83] a similar approach based on solving a small linear program is used. [Sha59, RS72a] evaluate all breakpoints of a one-dimensional piecewise linear function to solve the line location problem. They prove that their algorithm runs in polynomial time. Megiddo and Tamir [MT83] were the first to prove Med1 for the line location problem with vertical distance and they could give an $O(n \log^2 n)$ approach to solve the problem. Finally [Zem84] derived an algorithm that solves the problem in linear time, in two dimensions as well as in the higher dimensional case, by using the method based on Megiddo's linear time algorithms for linear programming, see [Meg83, Meg84].

Note that some of the above mentioned papers ([MN83, MT83, Zem84]) deal with the line location problem with vertical distance, since this problem arises while solving line location problems with rectangular distance. This leads us to another reason for studying the vertical distance, namely the definition of the rectangular metric l_1. As we will see in Section 3.1 (Lemma 3.1), the rectangular distance between a point $x = (x_1, x_2) \in \mathbb{R}^2$ and a line $l_{s,-1,b} \subset \mathbb{R}^2$ (with $s, b \in \mathbb{R}$) can be calculated as

$$l_1(x,l) = \min\{|a_{m1}s - a_{m2} + b|, \frac{1}{|s|}|a_{m1}s - a_{m2} + b|\},$$

and this can be rewritten as

$$l_1(x,l) = \min\{d_{ver}(x,l), \frac{1}{|s|}d_{ver}(x,l)\}.$$

Therefore many algorithms and results for d_{ver} can be directly transferred to the case of the rectangular distance l_1. Furthermore, many results for the vertical

distance can be extended to block norm distances (Theorem 3.3 for block norm distances, as done in [Sch96], or Algorithm 4) and even to distances derived from arbitrary norms (see Theorems 3.3, 8.2, and 8.3 as well as Algorithms 5 and 6). For the proofs of some other theorems, structural results concerning only the vertical distance are helpful. Theorem 4.1 can for example be derived using a result about line location with vertical distance (Lemma 4.2) and the proof of Theorem 6.5 needs Lemma 6.4 which deals also with d_{ver}.
As line location problems with d_{ver} are convex, the properties Med1, Med2, Cen1, and Cen2 can easily be derived in this case.

Lemma 2.1 ([MT83, Sha78, MN83]) *For d_{ver} the following criteria hold.*

Med1: *There exists a median line passing through two of the existing facilities.*

Cen1: *There exists a center line which is at maximum distance from three of the existing facilities.*

Cen2: *If the weights are all equal there exists a center line which is parallel to a facet of $conv(\mathcal{E}x)$.*

Proof: Med1 is shown in [MT83]. [MN83] gives a proof for Cen1; an earlier proof of both, Cen1 and Cen2, can be found in [Sha78]. (An explicit proof of Med1 and Cen1 will be given with the help of a dual interpretation of the problem in the next section.)

QED

Lemma 2.2 *For d_{ver} every median line l^* is pseudo-halving, i.e., Med2 holds for the vertical distance.*

Proof: Let the line $l_{s^*,-1,b^*} = \{(x_1, x_2) : x_2 = s^* x_1 + b^*\}$ be an optimal line. Let the slope s^* be fixed and consider the one-dimensional function

$$\bar{f}(b) = \sum_{m \in \mathcal{M}} w_m |s^* a_{m1} - a_{m2} + b|.$$

Since $l_{s^*,-1,b^*}$ is optimal we use

$$f(l_{s^*,-1,b}) = \bar{f}(b) \text{ for all } b \in \mathbb{R}$$

and conclude that b^* minimizes $\bar{f}$.
From Lemma 1.9 we know that b^* minimizes $\bar{f}$ if and only if b^* is a median of the weighted set $\{(a_{m2} - s^* a_{m1}, w_m)\}$. Furthermore, b^* is a median of $\{(a_{m2} - s^* a_{m1}, w_m)\}$ if and only if

$$\sum_{m: a_{m2} > s^* a_{m1} + b^*} w_m \leq \frac{1}{2} W \text{ and}$$
$$\sum_{m: a_{m2} < s^* a_{m1} + b^*} w_m \leq \frac{1}{2} W$$

which equivalently yields that the line $l_{s^*,-1,b^*}$ is pseudo-halving.

QED

2.2 A Dual Interpretation

In the following we will introduce a dual interpretation for line location problems with vertical distance which will be needed in Section 4.1 and in Chapter 6. Consider the following transformation T mapping points to non-vertical lines and vice versa.

Definition 16 *We define the following* **transformation T**.
For any point $x = (x_1, x_2) \in \mathbb{R}^2$ let $T(x)$ define a line

$$T(x) \quad := \quad l_{-x_1,-1,x_2} = \{(s,b) : b = -x_1 s + x_2\},$$

and for any non-vertical line $l_{s,-1,b}$ define a point

$$T(l_{s,-1,b}) \quad := \quad (s,b).$$

The space of the transformed points and lines will be called the *dual space*. In the following we will indicate points $X, Y, \ldots$ and lines L in the dual space by capital letters.
It can be easily checked that the transformation does not change the vertical distance between points and lines, as the following lemma confirms.

Lemma 2.3 *Let $x \in \mathbb{R}^2$ be a point and l be a line. Then*

$$d_{ver}(x,l) = d_{ver}(T(l), T(x)).$$

In particular, $x \in l \Leftrightarrow T(l) \in T(x)$.

Proof:

$$\begin{aligned} d_{ver}(x, l_{s,-1,b}) &= |x_1 s - x_2 + b| \\ &= d_{ver}\left((s,b), l_{-x_1,-1,x_2}\right) = d_{ver}(T(l), T(x)) \end{aligned}$$

QED

Therefore we conclude the following theorem.

Theorem 2.1 *The location of a line that minimizes the sum (the maximum) of weighted vertical distances to a given set of points $\mathcal{E}x = \{Ex_1, Ex_2, \ldots, Ex_M\}$ is equivalent to the location of a point minimizing the sum (the maximum) of weighted vertical distances to a given set of lines $\{T(Ex_1), T(Ex_2), \ldots, T(Ex_M)\}$, i.e.,*

- $1l/\mathbb{R}^2/\cdot/d_{ver}/\sum$ *and* $1/\mathbb{R}^2/\mathcal{E}x = \{T(Ex_1), \ldots, T(Ex_M)\}/d_{ver}/\sum$ *are equivalent, and,*

- $1l/\mathbb{R}^2/\cdot/d_{ver}/\max$ *and* $1/\mathbb{R}^2/\mathcal{E}x = \{T(Ex_1), \ldots, T(Ex_M)\}/d_{ver}/\max$ *are equivalent.*

Thus, in the dual space we are searching a point X minimizing the distance to a given set of lines. Formally, for points X in the dual space, the objectives f and g can be rewritten as follows.

$$f(X) = \sum_{m \in \mathcal{M}} w_m d_{ver}(X, T(Ex_m)) \text{ and}$$

$$g(X) = \max_{m \in \mathcal{M}} w_m d_{ver}(X, T(Ex_m)).$$

X minimizes one of these functions if and only if $T^{-1}(X)$ is an optimal line for the corresponding line location problem. The set of optimal solutions in the dual space is denoted by $\mathcal{X}^*_{med}$ and $\mathcal{X}^*_{cen}$, respectively.
Other transformations mapping points to lines and lines to points are often used in projective geometry. Similar transformations to the one given above that also transform points to lines and vice versa and that are not affecting the vertical distances between lines and points can be found in the literature of computational geometry and statistics, see e.g., [Ede87, SA95, CP95] and [KA97].

Example 2.1 *Consider a location problem with the following five existing facilities:*

$$\begin{aligned} Ex_1 &= (1, -1), \\ Ex_2 &= (-1, 1), \\ Ex_3 &= (-1, 2), \\ Ex_4 &= (0, 1), \text{ and} \\ Ex_5 &= (-\frac{1}{2}, -1). \end{aligned}$$

Figure 2.1 shows the set of existing facilities and the unique optimal lines l^*_{med} for the median problem and l^*_{cen} for the center problem, respectively. (In this example l^*_{med} and l^*_{cen} are parallel.) The transformed existing facilities $L_m = T(Ex_m), m = 1, \ldots, 5$ and the optimal solutions X^*_{med} and X^*_{cen} in the dual space are shown in Figures 2.2 and 2.3. How these solutions are obtained is shown in the next section.

2.2.1 The Median Problem

Consider the set of lines

$$\mathcal{K}^{med} := \{L_m = T(Ex_m) : m \in \mathcal{M}\}$$

that partitions the dual $\mathbb{R}^2$ into cells $C \in \mathcal{C}$, see Figure 2.2. In each cell, the sign of $(a_{m1}s - a_{m2} + b)$ does not change for all $m \in \mathcal{M}$, and thus the objective

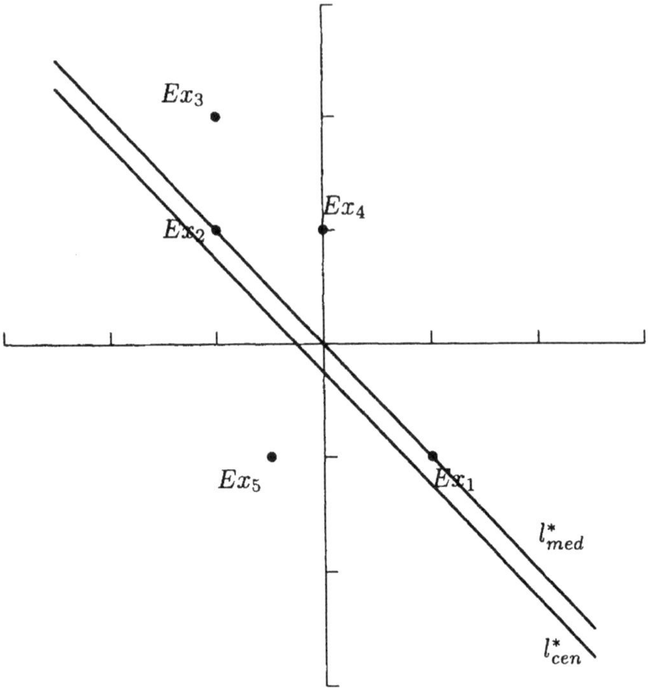

Figure 2.1. An example with five existing facilities and the unique optimal solutions l^*_{med} and l^*_{cen} for the median and the center problem, respectively.

function f of $1l/\mathbb{R}^2/\cdot/d_{ver}/\sum$ is linear on each cell. Since f is also convex we have a piecewise linear convex problem as introduced in Section 1.5 with a set of construction lines $\mathcal{K}^{med}$. In this case we have that $\mathcal{P}(\mathcal{K}^{med}) = \mathcal{P}(\mathcal{C}^{\partial})$. (Recall that $\mathcal{C}^{\partial}$ denotes the set of all cell facets).

We are now able to give the following proof of Med1 in Lemma 2.1: Since f is linear on each cell, we know that there exists an optimal solution which is an extreme point V of a cell (Theorem 1.3). Because all cell vertices lie on at least two different lines $T(Ex_m), T(Ex_k) \in \mathcal{K}^{med}$, we conclude that the line $T^{-1}(V)$ in the original space passes through the two points Ex_m and Ex_k (see Lemma 2.3). This means that there exists an optimal line for $1l/\mathbb{R}^2/\cdot/d_{ver}/\sum$ passing through at least two of the existing facilities.

Note that the number of cell vertices is quadratic in M. (Faster algorithms, such as described in Section 2.1, are possible since the set of candidates can be reduced to the breakpoints of the pointwise median of the lines $L_1, \ldots, L_M$.)

2.2.2 The Center Problem

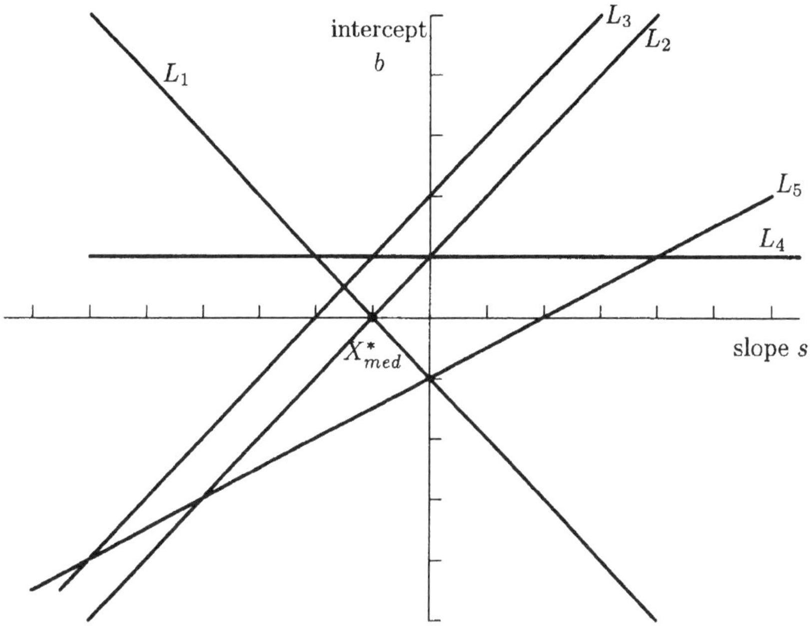

Figure 2.2. Construction lines and optimal solution for the median problem in the dual space.

For the center problem we can use the same theory as for the median problem because g is piecewise linear and convex, too. Only the cell structure differs. For describing the set of all cells $\mathcal{C}$ in this case we need the following definition.

Definition 17 *Let d be a distance and $A, B \subseteq \mathbb{R}^2$ be two sets.*

1. *The* **bisector** $b_{A,B}$ *of A and B with respect to the distance d is defined as*

$$b_{A,B} = \{x : d(x, A) = d(x, B)\}.$$

2. *Furthermore, let w_A, w_B be positive weights. The* **weighted bisector** $b^w_{A,B}$ *is then defined as*

$$b^w_{A,B} = \{x : w_A d(x, A) = w_B d(x, B)\}.$$

Note that this definition uses Definition 6 on page 7 in which we defined the distance d between two sets. If $A = \{x\}$ and $B = \{y\}$ are two distinct points and $d = l_2$ is the Euclidean distance, the perpendicular bisector $b_{x,y}$ is always a straight line. For the rectangular distance, e.g., this is not necessarily true; the bisector of two points $x, y \in \mathbb{R}^2$ may even contain two-dimensional pieces. For the vertical distance d_{ver} it can easily be verified that the bisector of two lines l', l'' is given by one or two lines passing through the intersection point of l' and l''. An example for the weighted bisector of two lines L_1 and L_2 with respect to the vertical distance is given in Figure 2.4.

Now let us come back to line location problems with vertical distance and center objective function. We first look at the unweighted center problem $1l/\mathbb{R}^2/w_m = 1/d_{ver}/\max$. Let us denote by U and L the upper and the lower envelope, respectively, of the set of lines $\{T(Ex_1), \dots, T(Ex_M)\}$ in the dual space. By definition of $g(x)$, L, and U

$$g(x) = \max_{m \in \mathcal{M}} d_{ver}(X, L_m) = \max\{d_{ver}(X, U), d_{ver}(X, L)\} \tag{2.1}$$

holds for all X in the dual space.
If we define the **mid-line** as the bisector of U and L, i.e.,

$$h^{Mid} = \{X : d_{ver}(X, U) = d_{ver}(X, L)\},$$

then for all $X \in h^{Mid}$ it follows that $d_{ver}(X, U) = d_{ver}(X, L) = g(X)$. Since U and L are piecewise linear functions, the bisector h^{Mid} of U and L can be calculated as follows. Let $\mathcal{H}_L$ and $\mathcal{H}_U$ denote the set of all first coordinates of breakpoints of L and U, respectively. Suppose that

$$\mathcal{H}_L \cup \mathcal{H}_U = \{z_1, z_2, \dots, z_Q\}$$

and let the indices be ordered such that $z_1 < z_2 < \dots < z_Q$. Then on each interval $I_q = [z_q, z_{q+1}], q = 1, 2, \dots, Q-1$, both functions U and L are linear. Hence the bisector of U and L in I_q can be calculated as the bisector of the two linear pieces $U_{|I_q}$ and $L_{|I_q}$. Consequently, h^{Mid} is piecewise linear with breakpoints whose first coordinates are in $\mathcal{H}_L \cup \mathcal{H}_U$ (see Figure 2.3 for an example).
In order to construct the set of cells $\mathcal{C}$ we define the following two sets of halflines:

$$h_z^l = \{X = (z, x_2) : X \text{ lies above } h_{Mid}\} \qquad \text{for all } z \in \mathcal{H}_L,$$
$$h_z^u = \{X = (z, x_2) : X \text{ lies below } h_{Mid}\} \qquad \text{for all } z \in \mathcal{H}_U.$$

We now can describe a set of construction lines for the center problem as

$$\mathcal{K}^{cen} = \{h : h \text{ is a linear piece of } h^{Mid}\} \cup \{h_{z_1}^l : z_1 \in \mathcal{H}_L\} \cup \{h_{z_2}^u : z_2 \in \mathcal{H}_u\}$$

and get again a cell partition of the plane (see Figure 2.3). From the properties of h^{Mid} and from Equation 2.1 we conclude that for each cell C there exists a $k \in \mathcal{M}$ such that

$$g(X) = d_{ver}(X, L_k) \text{ for all } X \in C.$$

Hence g is linear on the cells and our set of construction lines $\mathcal{K}^{cen}$ equals the set of all cell facets $\mathcal{C}^{\partial}$. The number of cell vertices equals the number Q of breakpoints of the mid-line h^{Mid} which is determined by the number of breakpoints of the upper and lower envelope U and L of the linear functions $L_m, m \in \mathcal{M}$. Since each line L_m can be a part of U and L in at most one interval, we get at most $M-1$ breakpoints of both U and L.

(In general, the number of pieces of the lower or upper envelope of a set of functions $\{h_1, h_2, \ldots, h_M\}$ with at most s intersections per pair of functions is given by the maximal length of the Davenport-Schinzel sequence $\lambda_s(M)$, compare [SA95, Kle97]. In our case $s = 1$, since two lines intersect in at most one point. Since $\lambda_1(M) = M$, this yields that L and U have at most M linear pieces, corresponding to $M-1$ breakpoints.)

Thus for h^{Mid} the number Q of breakpoints satisfies $Q \leq M-2$ such that the number of cell vertices for the unweighted center problem is of order $O(M)$.

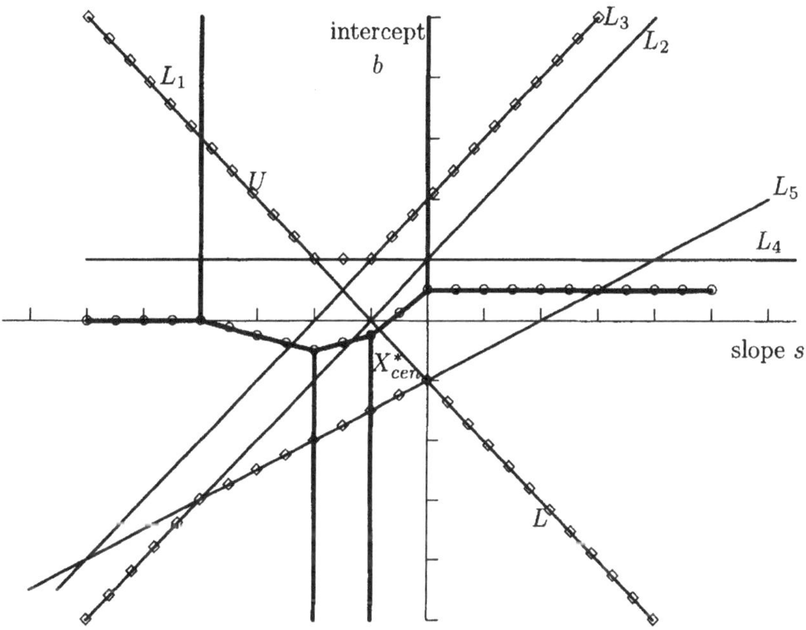

Figure 2.3. Construction lines and optimal solution for the unweighted center problem in the dual space. The mid-line h^{Mid} is marked by $\circ$, and $\diamond$ refers to U and L.

The objective $g(X)$ gives the maximum distance from the point X in the dual space to the set of lines $\{L_m : m \in \mathcal{M}\}$. For solving $1l/\mathbb{R}^2/\cdot/d_{ver}/\max$ we again know from Theorem 1.3 that there exists an optimal solution V which is a vertex of a cell. Thus in our case V is a breakpoint of h^{Mid} and its first coordinate is e.g. in $\mathcal{H}_L$. Since V is at maximum distance from U and L there exist three different lines L_{k1}, L_{k2}, and L_{k3} such that

$$\begin{aligned} g(V) &= d_{ver}(V,U) = d_{ver}(V,L_{k1}) \text{ and} \\ g(V) &= d_{ver}(V,L) = d_{ver}(V,L_{k2}) = d_{ver}(V,L_{k3}). \end{aligned}$$

Hence the corresponding line $T^{-1}(V)$ is at maximum distance from Ex_{k1}, Ex_{k2}, and Ex_{k3}. This proves Cen1 in Lemma 2.1 for the unweighted case.

One can also derive Cen2 using the dual interpretation of $1l/\mathbb{R}^2/\cdot/d_{ver}/\max$. To this end, we again remark that each breakpoint V of the mid-line corresponds to either a breakpoint of U or a breakpoint of L at this first coordinate, i.e., if $V = (s, b)$ is a breakpoint of h^{Mid} then there exists $V' = (s', b')$ with

$$V' \in L \cup U \text{ and } s' = s.$$

Since $s' = s$, the lines $l = T^{-1}(V)$ and $l' = T^{-1}(V')$ are parallel. To prove that l' contains a facet of $conv(\mathcal{E}x)$ we need a result which will be developed in Chapter 6. Using Lemma 6.3 we can transform $conv(\mathcal{E}x)$ to the dual space, i.e.,

$$\begin{aligned} T(conv(\mathcal{E}x)) &:= \{X : T^{-1}(X) \cap conv(\mathcal{E}x) \neq \emptyset\} \\ &= \{X : X \text{ lies between } L \text{ and } U,\} \end{aligned}$$

and we observe that V' is a vertex of $T(conv(\mathcal{E}x))$ and therefore $T^{-1}(V')$ contains a facet of $conv(\mathcal{E}x)$ (see Lemma 6.3 again). Summarizing, we can conclude that each line $T^{-1}(V)$ with a breakpoint V of h^{Mid} is parallel to a facet l' of $conv(\mathcal{E}x)$. Because there exists a breakpoint V^* minimizing g, this completes the proof.

Note that for all points $X \in h$ and for any $h \in \mathcal{K}^{cen}$ the line $T^{-1}(X)$ is at maximum distance from at least two of the existing facilities.

Finally we discuss the cell structure for the weighted case $1l/\mathbb{R}^2/\cdot/d_{ver}/\max$. First, we consider only two existing facilities corresponding to lines L_1 and L_2 with weights w_1 and w_2, respectively. The set of all points with the same weighted vertical distance to these two lines, i.e., the weighted bisector of L_1 and L_2 with respect to the vertical distance

$$b^w_{L_1,L_2} = \{X : w_1 d_{ver}(X, L_1) = w_2 d_{ver}(X, L_2)\}$$

consists in most cases of two lines l^1 and l^2 passing through the intersection point of L_1 and L_2. If $b^w_{L_1,L_2}$ consists only of one line l^1, this line lies between L_1 and L_2. In most cases, $b^w_{L_1,L_2}$ also consists of another line l^2 above or below both lines L_1 and L_2 (see Figure 2.4). Note that l^2 is a vertical line if and only if $w_1 = w_2$. If L_1 and L_2 are parallel and have equal weights, then no line l^2 exists. In the example of only two existing facilities, the plane is separated into at most four cells by the lines l^1 and l^2.
Now we turn to the case of M existing facilities. Note that in this case the objective function g still is piecewise linear and convex. This means that a cell partition $\mathcal{C}$ exists such that g is linear for all $C \in \mathcal{C}$. But instead of determining the set of facets of these cells $\mathcal{C}^\partial$, which in the case of the weighted center problem is denoted by $\mathcal{K}^{cen}_{weights}$, we define the following set of construction lines $\mathcal{H}$.

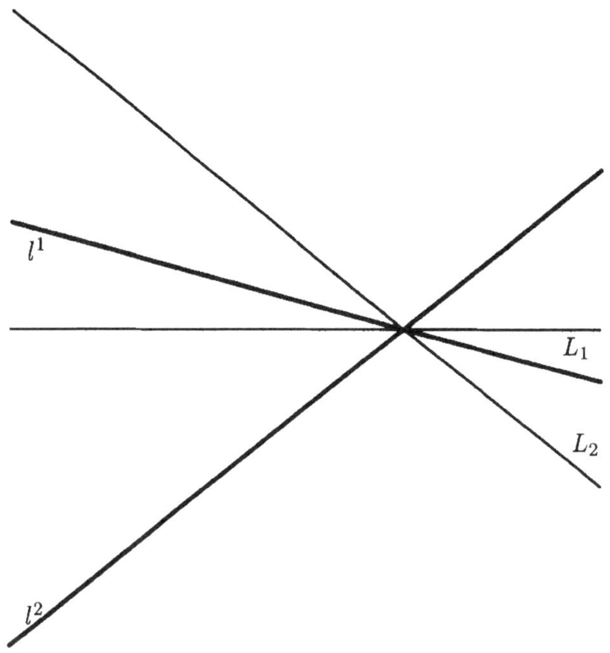

Figure 2.4. The two lines l^1, l^2 containing all points which have the same weighted vertical distance to L_1 and L_2 with weights $w_1 = 2, w_2 = 1$.

$$\mathcal{H} = \{l^1_{k,l}, l^2_{k,l} : k, l \in \mathcal{M}\}$$

with

$$\begin{aligned} l^1_{k,l} &= \{X = (s,b) : w_k d_{ver}(X, L_k) = w_l d_{ver}(X, L_l) \text{ and} \\ &\quad sgn(a_{k2} - sa_{k1} + b) = -sgn(a_{l2} - sa_{l1} + b)\} \\ l^2_{k,l} &= \{X = (s,b) : w_k d_{ver}(X, L_k) = w_l d_{ver}(X, L_l) \text{ and} \\ &\quad sgn(a_{k2} - sa_{k1} + b) = sgn(a_{l2} - sa_{l1} + b)\}. \end{aligned}$$

Let C be a cell defined by the above set of construction lines $\mathcal{H}$. Then there exists an $m^C \in \mathcal{M}$ such that $\max_{m \in \mathcal{M}} w_m d_{ver}(X, L_m) = w_{m^C} d_{ver}(X, L_{m^C})$ for all points $X \in C$. We conclude that g is linear on C, i.e., $\mathcal{H}$ is a set of construction lines for our problem, according to Definition 14 on page 24, and we have that

$$\begin{aligned} \mathcal{P}(\mathcal{K}^{cen}_{weights}) &= \{X : X \in h \text{ for an } h \in \mathcal{K}^{cen}_{weights}\} \\ &\subseteq \{X : X \in l \text{ for an } l \in \mathcal{H}\} \\ &= \mathcal{P}(\mathcal{H}). \end{aligned}$$

Equivalently, if $h \in \mathcal{K}^{cen}_{weights}$ is a line segment or a line or a halfline then there exist $p \in \{1,2\}$ and $k, l \in \mathcal{M}$ such that $h \subseteq l^p_{k,l}$.
Two examples for the real cell structure $\mathcal{K}^{cen}_{weights}$ for the weighted center problem are given in Figures 2.5 and 2.6. Note that for finding the optimal solution, it suffices to consider only cell vertices in the **mid-line** with

$$h^{Mid} = \{X : X \in h \text{ for } h \in \mathcal{K}^{cen}_{weights} \text{ with } h \subseteq l^1_{k,l} \text{ for } k, l \in \mathcal{M}\}.$$

Note that since f is convex, the cells are also convex. Furthermore, since for all line segments $s \in \mathcal{K}^{cen}_{weights}$ there exists a construction line $l \in \mathcal{H}$ with $s \subseteq l$ we get the following properties.

- For all $X \in h$ with $h \in h^p_{k,l}$ for $p \in \{1,2\}$ and $k, l \in \mathcal{M}$ we have that

$$g(V) \quad = \quad w_k d_{ver}(V, L_k) = w_l d_{ver}(V, L_l).$$

- For all cell vertices $V \in h^{Mid}$ there exist $k, l, m \in \mathcal{M}$ such that

$$\begin{aligned} g(V) &= w_k d_{ver}(V, L_k) \\ &= w_l d_{ver}(V, L_l) \\ &= w_m d_{ver}(V, L_m). \end{aligned}$$

From the second property we can conclude that Cen1 holds for line location problems with vertical distance, but looking at Figures 2.5 and 2.6 we observe that the result of Cen2 given for the unweighted case cannot be transferred to the weighted problem.

Example 2.2 *We use the same example as before (see Example 2.1) but with weights $w_m = 1$ for $m \in \{1,2,3,5\}$, and $w_4 = 3$. The set of cells C is given in Figure 2.5. Note that the mid-line has only two breakpoints which have to be evaluated in order to find the optimal solution. As Ex_4 has the largest weight of all existing facilities, $w_4 d_{ver}(X, Ex_4)$ determines the value of $g(X)$ when the second coordinate of X is sufficiently large or sufficiently small.*

Finally, another example with only 3 existing facilities is given.

Example 2.3 *Let the lines L_1, L_2, and L_3 be given as in Figure 2.6. Furthermore, let $w_1 = 3$, $w_2 = 3$ and $w_3 = 1$. Since the maximum weight occurs twice, we get vertical halflines.*

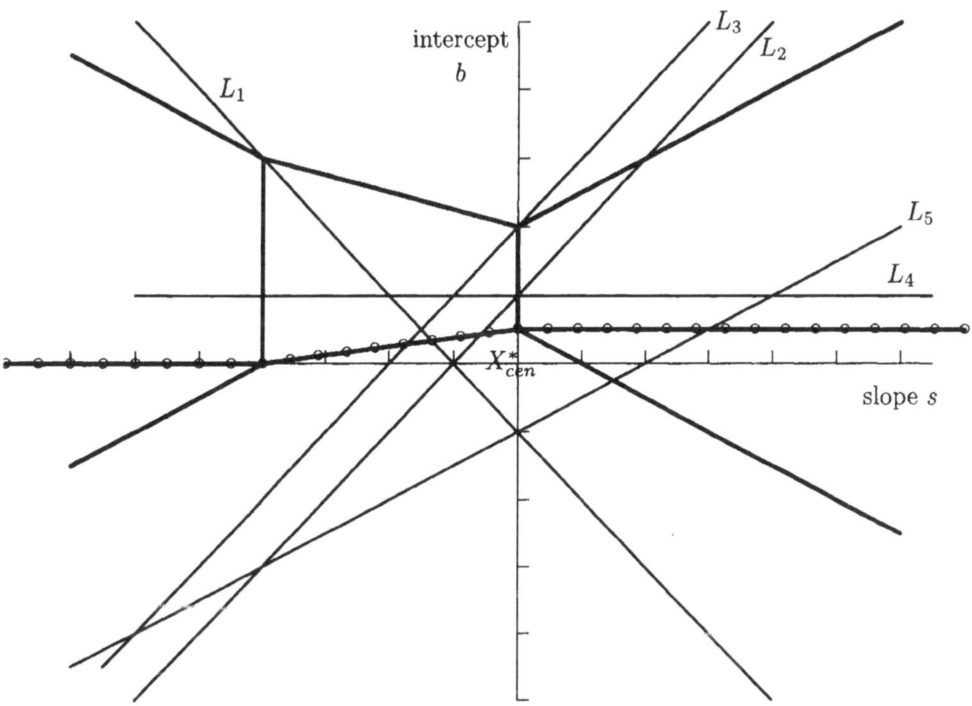

Figure 2.5. Cell structure and optimal solution for a weighted center problem in the dual space (cf. Example 2.2). The mid-line h^{Mid} is marked by $\circ$.

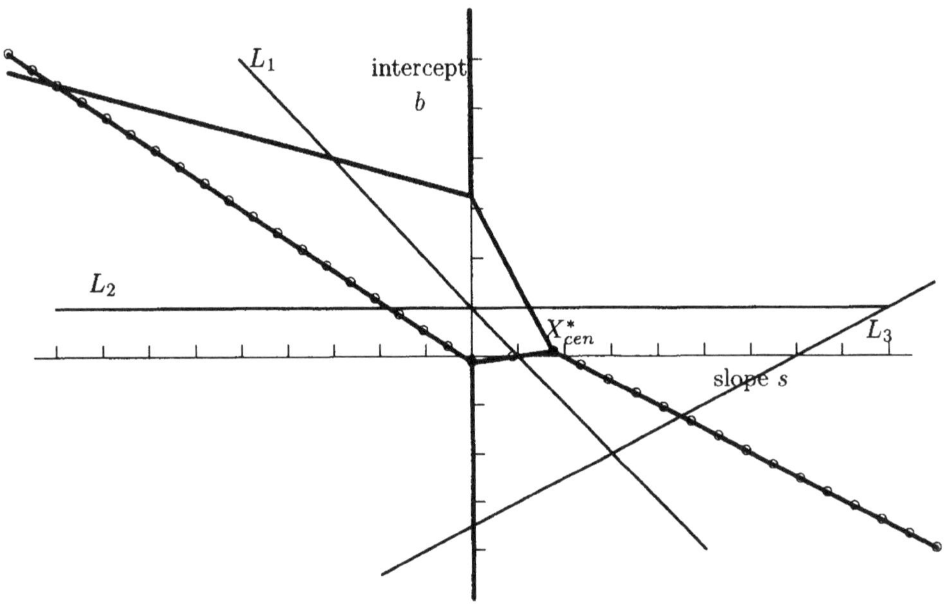

Figure 2.6. Cell structure and optimal solution for a weighted center problem in the dual space (cf. Example 2.3). The mid-line h^{Mid} is marked by $\circ$.

3
LOCATING LINES IN THE NORMED PLANE

3.1 Results for p-Norm Distances

In this section we discuss line location problems with p-norms, i.e., we consider problems of the type $(1l/\mathbb{R}^2/\cdot/l_p/\cdot)$. Note that none of the results given in this section is necessary to prove the results of the following sections. On the other hand, most of the line location problems discussed in the literature are dealing with p-norms (especially l_1 and l_2) such that this case deserves a section of its own. First, let us discuss the case $p = 1$.

3.1.1 The Rectangular Distance

We first give a formula for the rectangular distance l_1 between a point x and a non-vertical line l in $\mathbb{R}^2$.

Lemma 3.1 ([MN83, MT83, Mes95]) *Let $l = l_{s,-1,b}$ be a non-vertical line and $x \in \mathbb{R}^2$. Then the rectangular distance between x and l is given by*

$$l_1(x,l) = \min\{|a_{m1}s - a_{m2} + b|, \frac{1}{|s|}|a_{m1}s - a_{m2} + b|\}.$$

To rewrite Lemma 3.1 we calculate the horizontal distance between a point and a line, similar to the calculation of the vertical distance between a point and a line (see Section 2.1). For a non-vertical line $l_{s,-1,b}$ and a point x

$$d_{hor}(x, l_{s,-1,b}) = \frac{|x_1 s - x_2 + b|}{|s|}.$$

Between a vertical line $l_{-1,0,b}$ and a point x we get $d_{hor}(x, l_{-1,0,b}) = |x_1 - b|$. Therefore, the formula in Lemma 3.1 can be rewritten as

$$\begin{aligned} l_1(x,l) &= \min\{d_{ver}(x,l), d_{hor}(x,l)\} \\ &= \min\{d_{ver}(x,l), \frac{1}{|s|} d_{ver}(x,l)\}. \end{aligned}$$

Note that the smaller value of d_{hor} and d_{ver} can be found by investigating only the slope $s = -\frac{s_1}{s_2}$ of the line $l_{s_1,s_2,b}$. We have that

$$\begin{array}{lcll} d_{hor}(x,l) & < & d_{ver}(x,l) & \text{if } l = l_{s,-1,b} \text{ with } s > 1 \\ & & & \text{or if } l = l_{s,0,b} \text{ is a vertical line,} \\ d_{hor}(x,l) & = & d_{ver}(x,l) & \text{if } l = l_{1,-1,b} \text{ and} \\ d_{hor}(x,l) & > & d_{ver}(x,l) & \text{if } l = l_{s,-1,b} \text{ with } s < 1. \end{array}$$

To solve line location problems with l_1-distance we determine a line minimizing d_{ver} and a line minimizing d_{hor} and choose the one with smaller objective value. In Chapter 2 we have shown Med1, Med2, Cen1, and Cen2 for the vertical distance d_{ver}, see Lemma 2.1 and Lemma 2.2. According to Lemma 1.4 (on page 7) both Lemmas do hold also for the horizontal distance d_{hor} by rotating the coordinate system around 90^0. Therefore we can conclude:

Lemma 3.2 (see also [MT83, MN83, Sha78]) *For the rectangular distance l_1 the following criteria hold.*

Med1: *There exists a median line passing through two of the existing facilities.*

Med2: *All median lines are pseudo-halving.*

Cen1: *There exists a center line which is at maximum distance from three of the existing facilities.*

Cen2: *If the weights are all equal there exists a center line which is parallel to a facet of $conv(\mathcal{E}x)$.*

Proof: Med1 is shown in [MT83]. [MN83] gives a proof of Cen1; an earlier proof of both, Cen1 and Cen2, can be found in [Sha78]. They all use the analogous properties for the vertical distance (see Lemma 2.1) to show these results. Med2 can be derived directly from the property Med2 for the vertical distance (Lemma 2.2).

QED

Thus the algorithmic approaches for $1l/\mathbb{R}^2/\cdot/d_{ver}/\cdot$ described in Section 2.1 can be used to solve the line location problem with rectangular distance. For the median problem [MT83] gives an $O(M \log^2 M)$ approach, and [Zem84] solves the problem in linear time by using the method of Megiddo's linear time algorithms for linear programming, see [Meg83, Meg84]. The center problem with l_1 distance can be solved directly by the same linear programming methods.

3.1.2 The Chebyshev Distance

Using Lemma 3.2 and the transformation T described in Section 1.2 (see Lemma 1.2) we can transfer Med1, Med2, Cen1, and Cen2 also to the case of the l_∞ norm.

Lemma 3.3 *For the Chebyshev distance $d = l_\infty$ the properties Med1, Med2, Cen1, and Cen2 hold.*

Before we consider the general case of $p \in (1, \infty)$ we give a survey on the literature concerning Euclidean line location problems.

3.1.3 The Euclidean Distance

The distance between a point $x \in \mathbb{R}^2$ and a line $l = l_{s_1,s_2,b}$ is given by the length of the orthogonal projection of the point $x = (x_1, x_2)$ onto the line l. In [MN80] the distance from x to l has been calculated as

$$l_2(x,l) = \frac{|x_2 - sx_1 + b|}{\sqrt{s^2+1}},$$

where $l = l_{s,-1,b}$ is a non-vertical line. The formula given in [MT82] considers the case of a line $l = l_{s_1,s_2,0}$ passing through the origin. In this case we get

$$l_2(x,l) = \frac{|s_1x_1 + s_2x_2|}{\sqrt{s_1^2+s_2^2}}.$$

Later in this section we will develop a formula for the p-norm distance between a point and a line, which can be applied to all possible cases, for all $p \in (1, \infty)$.

For $1l/\mathbb{R}^2/\cdot/l_2/\sum$ [MN80] (and later also [MT82, LMW88, KM90]) showed that there always exists an optimal line passing through two of the given points. This was used in [MN80] to derive an $O(M^3)$ time algorithm. This result was improved in [MT83] to an $O(M^2 \log M)$ time algorithm, and [LC85] further improved this to $O(M^2)$. Finally, [YKII88] and, independently, [KM90] derived an $O(M^2)$ time approach, see also [KM93].
Much more is known for the special case with identical weights, i.e., for the planar unweighted case $1l/\mathbb{R}^2/w_m = 1/l_2/\sum$. For such line location problems [MN80] observed a second criterion: optimal lines have to be halving in the sense of Definition 9. Hence, for the unweighted situation purely combinatorial properties of the given point set are important. The following subquadratic bounds on the number $h(M)$ of halving lines are known:

$$h(M) \ < \ M^{\frac{3}{2}} \text{ (cf. [Lov71] and [ELSS73])}, \qquad (3.1)$$

$$h(M) \ < \ M^{\frac{3}{2}} \log^{-\frac{1}{100}} M \text{ (see [PSS92])}. \qquad (3.2)$$

In order to use these bounds to improve the time complexities mentioned above, it is necessary to implement the halving line procedure as described in [Lov71].

This procedure will briefly be explained in the following. Starting with a halving line l_1 that passes through two of the existing facilities (e.g., through Ex_1 and Ex_2), this line is rotated clockwise about Ex_2 until it hits some other point (e.g., Ex_3) to obtain $l_2 = aff(Ex_2, Ex_3)$. Then l_2 is rotated clockwise about Ex_3 to get l_3, and so on, until we return to the starting position. For odd M, all lines l_i are halving, and for even M, some of the lines l_i are halving and some are so-called $(\frac{M}{2}, \frac{M}{2} - 2)$-dividers, i.e., they separate the set of existing facilities into a set of cardinality $\frac{M}{2}$ and into a set of cardinality $\frac{M}{2} - 2$. (Hence all lines l_i are pseudo-halving.) Using the data structure given in [OvL81], the rotation procedure of [Lov71] can be implemented in $O(h(M) \log^2 M)$ time, because the number of $(\frac{M}{2}, \frac{M}{2} - 2)$–dividers in the even case has asymptotically the same upper bound as the number of halving lines. This led [YKII88] to an $O(M^{\frac{3}{2}} \log^2 M)$ time algorithm by using (3.1), and using (3.2) this can be improved to an $O(M^{\frac{3}{2}} \log^{2-\frac{1}{100}} M)$ time approach, see [KM93]. The question about a time optimal algorithm remains open, yet. The known lower bound is $\Omega(M \log M)$, proven in [YKII88] by reduction from the so-called uniform gap on a circle problem.

The search for center lines with respect to the Euclidean distance is also called the orthogonal L_∞ approximation problem. [MN83] showed that each optimal line for the center problem is at maximum (weighted) distance from at least three of the existing facilities. On the base of this result and followed by various independent approaches (compare [HII+93] and [KM93] for an exhaustive list of contributions), [Dor84] and [HT85] presented an $O(M \log M)$ time algorithm for the unweighted problem. Its optimality was shown in [LW86]. An $O(M^2 \log M)$ time algorithm for the weighted case was also given in [LW86]. The approach in [Ede85] runs in $O(M \log M)$ time and was shown to be optimal for $1l/\mathbb{R}^2/\cdot/l_2/\max$.

Summarizing the discussion above we can state the following lemma.

Lemma 3.4 (many papers) *For the Euclidean distance the criteria Med1, Med2, Cen1, and Cen2 hold.*

Proof: At least three different proofs for Med1 can be found in [MN80], [MT83], [KM90], and [LMW88]. The first three papers also prove Med2. Proofs for Cen1 and Cen2 can be found, e.g., in [MN83].

QED

3.1.4 *Other p-Norm Distances*

In the following we will discuss the case of general p-norms, i.e., $p \in (1, \infty)$. First we derive a formula to determine the p-norm distance between a point $x = (x_1, x_2) \in \mathbb{R}^2$ and a line $l = l_{s_1,s_2,b}$.

Lemma 3.5 *Let $p \in [1, \infty), x \in \mathbb{R}^2$ and l a line. Then*

$$l_p(x, l) = \begin{cases} \frac{|sx_1 - x_2 + b|}{\sqrt[p]{\left(1 + |s| \sqrt[p-1]{|s|}\right)^{p-1}}} & \text{if } l = l_{s,-1,b} \text{ is a non-vertical line,} \\ |x_1 + \frac{b}{s_1}| & \text{if } l = l_{s_1,0,b} \text{ is a vertical line.} \end{cases}$$

Proof: This formula can be derived by solving the following minimization problem:

$$\min_{y \in l} l_p(x, y).$$

Case 1 : For $s_2 = 0$ we consider a line $l_{s_1,0,b} = \{(y_1, y_2) : y_1 = -\frac{b}{s_1}\}$. Then the minimization problem reduces to

$$\min_{y_2 \in \mathbb{R}} \sqrt[p]{\left|x_1 + \frac{b}{s_1}\right|^p + |x_2 - y_2|^p}$$

such that we get $y^* = (-\frac{b}{s_1}, x_2)$ as the minimizer and

$$l_p(x, l_{s_1,0,b}) = l_p(x, y^*) = \left|x_1 + \frac{b}{s_1}\right|$$

as the distance between x and the line. Note that $l_p(x, l) = d_{hor}(x, l)$ in this case.

Case 2 : Now we can assume that $s_2 \neq 0$ and therefore, without loss of generality, $s_2 = -1$. Furthermore, let us first assume that $b = 0$. Thus we consider a line $l_{s,-1,0} = \{(y_1, y_2) : y_2 = sy_1\}$. The minimization problem can be written as

$$\min_{y_1 \in \mathbb{R}} \sqrt[p]{|x_1 - y_1|^p + |x_2 - sy_1|^p}.$$

Equivalently we can minimize the one-dimensional function

$$h(y_1) = |x_1 - y_1|^p + |x_2 - sy_1|^p,$$

which is convex and therefore has a unique minimum. This minimum lies in the interval $[x_1, \frac{x_2}{s}]$ or in $[\frac{x_2}{s}, x_1]$, because for y_1 smaller than the values in the interval, h is decreasing, and on the right hand side of the interval it is an increasing function. To get rid of the absolute values we distinguish two cases.

$x_1 \leq \frac{x_2}{s}$: Then for $y_1 \in [x_1, \frac{x_2}{s}]$ we get

$$h(y_1) = \begin{cases} (-x_1 + y_1)^p + (x_2 - sy_1)^p & \text{if } s \geq 0 \\ (-x_1 + y_1)^p + (-x_2 + sy_1)^p & \text{if } s < 0. \end{cases}$$

$x_1 > \frac{x_2}{s}$: Then for $y_1 \in [\frac{x_2}{s}, x_1]$ we get

$$h(y_1) = \begin{cases} (x_1 - y_1)^p + (-x_2 + sy_1)^p & \text{if } s \geq 0 \\ (x_1 - y_1)^p + (x_2 - sy_1)^p & \text{if } s < 0. \end{cases}$$

In these intervals, h is differentiable. Therefore calculating h' and solving the equation $h'(y_1) = 0$ leads to the unique minimizer $y^* = (y_1^*, y_2^*)$ with

$$y_1^* = \frac{x_1 + sgn(s) \sqrt[p-1]{|s|} x_2}{1 + |s| \sqrt[p-1]{|s|}},$$

$$y_2^* = sy_1^*.$$

Defining $\mathfrak{s} = sgn(s) \sqrt[p-1]{|s|}$, the distance between the line $l_{s,-1,0}$ and the point $x = (x_1, x_2)$ is finally given by

$$\begin{aligned} l_p(x, l_{s,-1,0}) &= l_p(x, y^*) \\ &= \sqrt[p]{|x_1 - y_1^*|^p + |x_2 - sy_1^*|^p} \\ &= \left[\left(\frac{|x_1 + s\mathfrak{s}x_1 - x_1 - \mathfrak{s}x_2|^p}{1 + s\mathfrak{s}} \right)^p \right. \\ &\quad \left. + \left(\frac{|x_2 + s\mathfrak{s}x_2 - sx_1 - s\mathfrak{s}x_2|^p}{1 + s\mathfrak{s}} \right)^p \right]^{\frac{1}{p}} \\ &= \frac{1}{1 + s\mathfrak{s}} \sqrt[p]{|sx_1 - x_2|^p (s\mathfrak{s} + 1)} \\ &= \frac{|sx_1 - x_2|}{\sqrt[p]{\left(1 + |s| \sqrt[p-1]{|s|}\right)^{p-1}}}. \end{aligned}$$

For $b \neq 0$ we use the following translation:

$$\begin{aligned} l_p((x_1, x_2), l_{s,-1,b}) &= l_p((x_1, x_2 - b), l_{s,-1,0}) \\ &= \frac{|sx_1 - x_2 + b|}{\sqrt[p]{\left(1 + |s| \sqrt[p-1]{|s|}\right)^{p-1}}}, \end{aligned}$$

which finally proves the result for all possible lines in $\mathbb{R}^2$.

QED

Lemma 3.6 *For the median problem with distance l_p, $p \in (1, \infty)$ the following holds:*

All median lines are passing through two of the existing facilities, i.e., the weak incidence property and the strong incidence property hold for $1l/\mathbb{R}^2/\cdot/l_p/\sum$ with $p \in (0, 1)$.

Proof: This proof is based on the concavity of the objective function while rotating the line about one of the existing facilities. It is a generalization of the proof for the Euclidean distance which was given by [MN80]. Note that in Theorem 3.3 (in the next section) we will prove a more general version of the weak incidence property, while a more general version of the strong incidence property will be given in Theorem 4.1.

1. Non-vertical lines: Using Lemma 3.5 we get the following one-dimensional minimization problem for any given slope $s \neq \infty$:

$$\min_{b \in \mathbb{R}} \sum_{m \in \mathcal{M}} w'_m |a'_m + b|$$

with

$$\begin{aligned} w'_m &= \frac{w_m}{\sqrt[p]{\left(1 + |s|^{\sqrt[p-1]{|s|}}\right)^{p-1}}} \\ a'_m &= s a_{m1} - a_{m2}. \end{aligned}$$

For this problem we know from Lemma 1.9 that the set of optimal solutions is given by $median\{(-a'_m, w'_m) : m \in \mathcal{M}\}$, i.e., there exists a $k \in \mathcal{M}$ such that

$$b^* = -a'_k = -s a_{k1} + a_{k2}.$$

We therefore conclude that there exists an optimal line that passes through one of the existing facilities and, without loss of generality, we assume that this existing facility is $Ex_k = 0$. Therefore we will only investigate lines $l = l_{s,-1,0}$ passing through the origin. That leads to $b = 0$ and simplifies the objective function to

$$\min_{s \in \mathbb{R}} \sum_{m \in \mathcal{M}} w_m \frac{|s a_{m1} - a_{m2}|}{\sqrt[p]{\left(1 + |s|^{\sqrt[p-1]{|s|}}\right)^{p-1}}}.$$

Let us denote this one-dimensional objective function by $f(s)$ and assume that s^* is optimal. In order to prove Med1 we have to show that the line $l_{s^*,-1,0}$ passes through one of the points Ex_m (since the origin is the first existing facility and it already lies on l^*). Because $l_p(0, l_{s^*,-1,0}) = 0$, the origin can be neglected in the following, i.e., we can assume that $Ex_m \neq 0$ for all $m \in \mathcal{M}$. Now assume that Med1 does not hold, i.e.,

$$|s^* a_{m1} - a_{m2}| \neq 0 \text{ for all } m \in \mathcal{M}.$$

Because of the continuity of $|s a_{m1} - a_{m2}|$ there exists an open interval $U(s^*) = (s^* - \epsilon, s^* + \epsilon)$ such that

$$|s a_{m1} - a_{m2}| \neq 0 \text{ for all } m \in \mathcal{M} \text{ and for all } s \in U(s^*).$$

1.1 Non-vertical and non-horizontal lines Here we need the assumption that $s^* \neq 0$, since then we find an $\epsilon > 0$ such that $0 \notin U(s^*)$ and f is differentiable for all $s \in U(s^*)$. Hence we can determine the first and the second derivative of f. Therefore we define

$$\begin{aligned} \mathcal{M}^+ &= \{m \in \mathcal{M} : sa_{m1} - a_{m2} > 0\}, \\ \mathcal{M}^- &= \{m \in \mathcal{M} : sa_{m1} - a_{m2} < 0\}, \end{aligned}$$

and from our assumptions we can conclude that $\mathcal{M} = \mathcal{M}^+ \cup \mathcal{M}^-$. To simplify the notation we introduce the constants A_2 and A_1 as

$$\begin{aligned} A_1 &= \sum_{m \in \mathcal{M}^-} w_m a_{m1} - \sum_{m \in \mathcal{M}^+} w_m a_{m1}, \\ A_2 &= \sum_{m \in \mathcal{M}^-} w_m a_{m2} - \sum_{m \in \mathcal{M}^+} w_m a_{m2}. \end{aligned}$$

With this notation f, f' and f'' can be rewritten as

$s^* > 0$:

$$\begin{aligned} f(s) &= \frac{1}{\left(1 + s^{\frac{p}{p-1}}\right)^{\frac{p-1}{p}}} (A_2 - sA_1), \\ f'(s) &= -\left(1 + s^{\frac{p}{p-1}}\right)^{\frac{1-2p}{p}} (s^{\frac{1}{p-1}} A_2 + A_1), \\ f''(s) &= \frac{\left(1 + s^{\frac{p}{p-1}}\right)^{\frac{1-3p}{p}} s^{\frac{1}{p-1}}}{p-1} \left(2A_2 s^{\frac{1}{p-1}} \left(p - 1 - \frac{1}{2} s^{-\frac{p}{p-1}}\right) + A_1(2p-1)\right) \end{aligned}$$

$s^* < 0$:

$$\begin{aligned} f(s) &= \frac{1}{\left(1 + (-s)^{\frac{p}{p-1}}\right)^{\frac{p-1}{p}}} (A_2 - sA_1), \\ f'(s) &= -\left(1 + (-s)^{\frac{p}{p-1}}\right)^{\frac{1-2p}{p}} ((-s)^{\frac{1}{p-1}} (-A_2) + A_1), \\ f''(s) &= \frac{\left(1 + (-s)^{\frac{p}{p-1}}\right)^{\frac{1-3p}{p}} (-s)^{\frac{1}{p-1}}}{p-1} \left(2A_2 (-s)^{\frac{1}{p-1}} \left(p - 1 - \frac{1}{2} (-s)^{-\frac{p}{p-1}}\right)\right. \\ &\qquad \left. - A_1(2p-1)\right). \end{aligned}$$

Now we want to find out if there exists a minimum in the interval $U(s^*)$. If not, s^* cannot be minimal which is a contradiction to our assumption.
For $s^* > 0$ we have that the equation $f'(s) = 0$ has only a solution s^0 if $A_2 A_1 < 0$, namely

$$s^0 = \left(-\frac{A_1}{A_2}\right)^{p-1}.$$

For $s^* < 0$ a solution of $f'(s) = 0$ exists only if $A_2 A_1 > 0$. In this case it is given by

$$s^0 = -\left(\frac{A_1}{A_2}\right)^{p-1}.$$

Now we calculate the sign of $f''(s^0)$. Therefore we define

$$C(s) = \frac{\left(1 + |s|^{\frac{p}{p-1}}\right)^{\frac{1-3p}{p}}}{p-1}.$$

Some algebraic calculations give the following values of $f''(s^0)$.

Case I: $s^* > 0$:

$$f''(s^0) = \frac{C(s)}{A_2 A_1^{p-2}} \left(-A_1^p + A_2^p(-1)^{p-1}\right),$$

and using that $A_2 A_1 > 0$ we get

$$\begin{array}{rcrcl} p \text{ is even} & \Longrightarrow & sgn(f''(s^0)) & = & -sgn(A_2), \\ p \text{ is odd} & \Longrightarrow & sgn(f''(s^0)) & = & sgn(A_2 A_1) sgn(A_2^p - A_1^p) \\ & & & = & sgn(A_1 - A_2) = sgn(-A_2). \end{array}$$

Case II: $s^* < 0$:

$$f''(s^0) = \frac{C(s)}{A_2 A_1^{p-2}} \left(-A_1^p - A_2^p\right).$$

In a similar way, using that $A_2 A_1 < 0$, we get

$$sgn(f''(s^0)) \quad = \quad -sgn(A_2).$$

To complete this part of the proof we note that

$$f(s^*) > 0 \Longrightarrow A_2 - s^* A_1 > 0.$$

Suppose that $sgn(A_2) \leq 0$. Then $A_2 \leq 0$ and, since not both A_2 and A_1 can be zero, we get

$s^* > 0 \Longrightarrow A_1 \geq 0$, i.e., $A_2 - s^* A_1 < 0$ and

$s^* < 0 \Longrightarrow A_1 \leq 0$, i.e., $A_2 - s^* A_1 < 0$,

which leads to a contradiction in both cases. Finally we conclude

$$sgn(f''(s^0)) = -sgn(A_2) < 0,$$

and therefore there is no minimum in the open interval U, a contradiction to s^* being the optimal slope.

1.2 Horizontal lines: The case $s^* = 0$ still remains open, because f is not differentiable at $s = 0$. The objective value for $s^* = 0$ is given by

$$f(0) = \sum_{m \in \mathcal{M}} w_m |a_{m2}|.$$

For $s \in \mathbb{R}$ we rewrite f as

$$\sum_{m \in \mathcal{M}} w_m |s a_{m1} - a_{m2}| = \sum_{m \in \mathcal{M}} w_m |a_{m1}| \left| s - \frac{a_{m2}}{a_{m1}} \right|.$$

Then the set of minimizers of this function is given by the median of the values $\frac{a_{m2}}{a_{m1}}$ with weights $w'_m = |a_{m1} w_m|$ (using Lemma 1.9 again). Thus there exists a median

$$\tilde{s} = \frac{a_{k2}}{a_{k1}} \text{ for some } k \in \mathcal{M}$$

with $\tilde{s} \neq s^*$, since for s^* we have that $|s^* a_{m1} - a_{m2}| \neq 0$ for all $m \in \mathcal{M}$. For the slope $\tilde{s}$ being a minimizer of $\sum_{m \in \mathcal{M}} w_m |s a_{m1} - a_{m2}|$, we conclude

$$\sum_{m \in \mathcal{M}} w_m |\tilde{s} a_{m1} - a_{m2}| \leq \sum_{m \in \mathcal{M}} w_m |a_{m2}|.$$

Furthermore, $|\tilde{s}| > 0$ yields

$$\frac{1}{\left(1 + |\tilde{s}|^{\frac{p}{p-1}}\right)^{\frac{p-1}{p}}} < 1.$$

Summarizing we get

$$\begin{aligned} f(\tilde{s}) &= \frac{1}{\left(1 + |\tilde{s}|^{\frac{p}{p-1}}\right)^{\frac{p-1}{p}}} \sum_{m \in \mathcal{M}} w_m |\tilde{s} a_{m1} - a_{m2}| \\ &< 1 \sum_{m \in \mathcal{M}} w_m | - a_{m2}| \\ &= f(0), \end{aligned}$$

which is a contradiction to the optimality of s^*.

2. Vertical lines: All l_p norms are invariant with respect to the rotation about 90^0, see Lemma 1.3. Hence we can use the result for horizontal lines. This tells us that no horizontal line is optimal if it is not passing through two of the existing facilities. The same holds for vertical lines by Lemma 1.3.

More precisely, also for vertical lines $l_{-1,0,b}$ we can choose b such that $l_{-1,0,b}$ passes through one of the existing facilities, and thus we can assume $b = 0$. For $l_{-1,0,0}$ the objective value is given as

$$f(l_{1,0,0}) = \sum_{m \in \mathcal{M}} w_m |a_{m1}|.$$

In Case 1.2 we showed that the sum of all second coordinates cannot be optimal and the symmetry of Lemma 1.3 then yields that also the sum of all first coordinates cannot be optimal.

QED

Similar methods can be applied to prove Cen1 for the l_p distance functions. This is omitted here, because in the next section we will confirm all four properties Med1, Med2, Cen1, and Cen2 for all distances derived from norms. But note again that the above proof gives a much stronger result than Med1: We proved that *all* optimal lines are passing through two of the existing facilities, if the distance measure is derived from a p-norm, $p \in (1, \infty)$. For arbitrary norms this strong incidence property will be discussed in Chapter 4.

3.2 Results for Arbitrary Norms

It was shown in [MT82] that for general r it is NP-hard to decide if there exist r straight lines $l_1, \ldots, l_r$ in the plane such that

$$\mathcal{E}x \subseteq \bigcup_{i=1,\ldots,r} l_i,$$

where $\mathcal{E}x$ is a given set of M points (the so called *line cover problem*). Consequently, it is NP-hard to locate r lines in the plane such that the sum of distances (or the maximum distance) between the existing facilities and the set of lines is minimized, i.e., both problems $r\ l/\mathbb{R}^2/\cdot/\gamma/\sum$ and $r\ l/\mathbb{R}^2/\cdot/\gamma/\max$ are NP-hard if the distance measure is derived from a norm. In the following we will prove that for $r = 1$ this problem is polynomially solvable. Some of the results of this section have already been published by the author, see [Sch98a]. In this section we do not use any of the results of Section 3.1 about line location problems with p-norms, whereas the results of Chapter 2 derived for the case of vertical distance play an important role in the following.

We start noting that, in order to determine the distance between a point x and a line l, we can dilate the unit ball around x until it touches the line.

Lemma 3.7 *Let d be a distance derived from a norm with unit ball B. Then for any line l and any point x in the plane*

$$d(x, l) = \min\{|\lambda| : (x + \lambda B) \cap l \neq \emptyset\}.$$

Proof:

$$\begin{aligned} d(x, l) &= \min_{y \in l} d(x, y) \\ &= \min_{y \in l} \min\{|\lambda| : y - x \in \lambda B\} \\ &= \min_{y \in l} \min\{|\lambda| : y \in \lambda B + x\} \\ &= \min\{|\lambda| : \exists y \in l \text{ such that } y \in \lambda B + x\} \\ &= \min\{|\lambda| : (x + \lambda B) \cap l \neq \emptyset\}. \end{aligned}$$

QED

Let d be any metric derived from a norm. Recall that the problems that we consider in this section can be stated as follows.
Find a line l minimizing

$$f(l) = \sum_{m \in \mathcal{M}} w_m d(Ex_m, l)$$

or

$$g(l) = \max_{m \in \mathcal{M}} w_m d(Ex_m, l),$$

respectively.
In the classification scheme we can describe these problems by $1l/\mathbb{R}^2/\cdot/\gamma/\sum$ and $1l/\mathbb{R}^2/\cdot/\gamma/\max$, respectively. For confirming Med1, Med2, Cen1, and Cen2 with respect to these location problems we first have to define another location problem, which can be easily solved.

Definition 18 *Let $t \in \mathbb{R}^2$ be a given direction. For two points x, y in the plane define the* **t-distance** $d_t(x, y)$ *by*

$$d_t(x, y) := \gamma_t(y - x),$$

where

$$\gamma_t(x) := \begin{cases} |\alpha| & \text{if } x = \alpha t \\ \infty & \text{else.} \end{cases}$$

Note that γ_t may be infinite and therefore does not define a norm. Nevertheless, for all $x, y \in \mathbb{R}^2$ and all $\alpha \in \mathbb{R}$, γ_t satisfies

$$\begin{aligned} &\gamma_t(x) \geq 0 \\ &\gamma_t(x) = 0 \iff x = 0 \\ &\gamma_t(\lambda x) = |\lambda| \gamma_t(x) \text{ and} \\ &\gamma_t(x + y) \leq \gamma_t(x) + \gamma_t(y). \end{aligned}$$

In the following we determine the t-distance between a point x and a line l.

Lemma 3.8 *For $t \in \mathbb{R}^2, x \in \mathbb{R}^2$ and any line $l \subset \mathbb{R}^2$*

$$d_t(x, l) := \min\{|\lambda| : x + \lambda t \in l\},$$

where $\min \emptyset := \infty$.

Proof: Using the definition of $d_t(x, l)$, namely $d_t(x, l) := \min_{y \in l} d_t(x, y)$, we can prove Lemma 3.8 in the same way as Lemma 3.7, but with

$$B := \{x : \gamma_t(x) \leq 1\} = \{x = \lambda t : |\lambda| \leq 1\}.$$

QED

If for a given point $x \in \mathbb{R}^2$ and a direction $t \in \mathbb{R}^2$

$$l = l^{x,t} = \{y \in \mathbb{R}^2 : y = x + \lambda t, \lambda \in \mathbb{R}\}$$

denotes the line through the point x in the direction $t \in \mathbb{R}^2$, and if $\|t\|$ denotes the Euclidean length of t, we get (see Figure 3.1)

$$d_t(x,l) := \begin{cases} \frac{l_2(x,y)}{\|t\|} & \text{if } l \cap l^{x,t} = \{y\} \\ 0 & \text{if } l = l^{x,t} \\ \infty & \text{if } l \cap l^{x,t} = \emptyset. \end{cases} \tag{3.3}$$

Note that $d_t(x,l) < \infty$ for all $x \in \mathbb{R}^2$ if and only if t and l do not have the same slope. (Recall that the slope s_t of a vector $t = (t_1, t_2) \in \mathbb{R}^2$ is defined by $s_t = \frac{t_2}{t_1} \in \mathbb{R} \cup \{\infty\}$.)

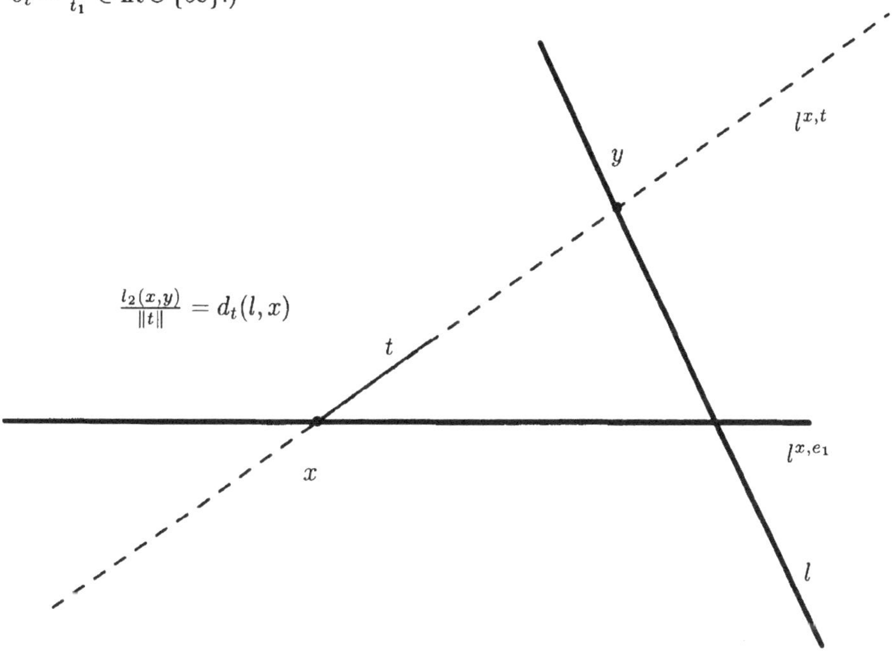

Figure 3.1. Illustration of Lemma 3.8.

For example, the length of the horizontal line segment from x to l is given by $d_{e_1}(x,l) = d_{hor}(x,l)$ and the length of the vertical line segment is given by $d_{e_2}(x,l) = d_{ver}(x,l)$, where e_1 and e_2 are the two unit vectors of $\mathbb{R}^2$. This shows that the t-distance is a generalisation of d_{ver} and d_{hor}.

We can now define the following two location problems $1l/\mathbb{R}^2/\cdot/d_t/\sum$ and $1l/\mathbb{R}^2/\cdot/d_t/\max$: For a given $t \in \mathbb{R}^2$ find a line l minimizing

$$f(l) = \sum_{m \in \mathcal{M}} w_m d_t(Ex_m, l)$$

or

$$g(l) = \max_{m \in \mathcal{M}} w_m d_t(Ex_m, l),$$

respectively.

The next lemma shows how to find an optimal line l^* if the distance is of the type d_t.

Lemma 3.9 *Let $p, q \in \mathbb{R}^2$ and D be a linear transformation with*

1. $D(p) = q$
2. $det(D) \neq 0$.

Then for all $x \in \mathbb{R}^2$

$$d_q(D(x), D(l)) = d_p(x, l),$$

where $D(l) := \{D(y) : y \in l\}$.

Note that such a transformation D exists whenever $p \neq 0$ and $q \neq 0$.

Proof: To prove the result we show first that

$$d_q(D(x), D(y)) = d_p(x, y) \text{ for points } x, y \in \mathbb{R}^2.$$

Case 1: $d_p(x, y) = \overline{\alpha} < \infty$: Then $x - y = \alpha p$ with $|\alpha| = \overline{\alpha}$, and we get

$$\begin{aligned} d_q(D(x), D(y)) &= \gamma_q(D(y) - D(x)) \\ &= \gamma_q(D(y - x)) \\ &= \gamma_q(D(\alpha p)) \\ &= \gamma_q(\alpha D(p)) \\ &= \gamma_q(\alpha q) = |\alpha| = \overline{\alpha}. \end{aligned}$$

Case 2: $d_p(x, y) = \infty$: Then we know that $x - y$ and p are linearly independent, and thus also $D(x - y)$ and $D(p)$ are linearly independent (because of $\det(D) \neq 0$), and we get

$$d_q(D(x), D(y)) = \infty.$$

We conclude that for a line l and any point x

$$\begin{aligned} d_q(D(x), D(l)) &= \min_{y \in l} d_q(D(x), D(y)) \\ &= \min_{y \in l} d_p(x, y) = d_p(x, l). \end{aligned}$$

QED

We can use Lemma 3.9 to transform $1l/\mathbb{R}^2/\cdot/d_t/\sum$ and $1l/\mathbb{R}^2/\cdot/d_t/\max$ to the corresponding line location problems with vertical distance described in Chapter 2. As a consequence we get the following theorem.

Theorem 3.1 *For all distances d_t the following criteria hold.*

Med1: *: There exists a median line passing through two of the existing facilities.*

Med2: *All median lines are pseudo-halving.*

Cen1: *There exists a center line which is at maximum distance from three of the existing facilities.*

Cen2: *If the weights are all equal, there exists a center line which is parallel to a facet of conv($\mathcal{E}x$).*

Proof: Using Lemma 3.9 with $p = t$ and $q = e_2$ we get

$$d_{e_2}(D(x), D(l)) = d_t(x, l) \text{ for all } x \in \mathbb{R}^2. \tag{3.4}$$

For the objective functions $f(l)$ and $g(l)$ it follows that

$$l \text{ is optimal for } d_t \text{ and } \mathcal{E}x \Longleftrightarrow D(l) \text{ is optimal for } d_{ver} \text{ and } D(\mathcal{E}x).$$

For d_{ver} we know from Lemma 2.2 and Lemma 2.1 that Med1, Med2, Cen1, and Cen2 are true. It remains to show that these properties are not affected by applying the transformation D. Since D is a homeomorphism we get for any set $A \subseteq \mathbb{R}^2$ that

$$D(\partial(A)) = \partial(D(A)). \tag{3.5}$$

Med1: Since D is a linear transformation,

$$Ex_m \in l \Longleftrightarrow D(Ex_m) \in D(l).$$

This means that $D(l)$ passes through at least two different points $D(Ex_k)$ and $D(Ex_l)$ if and only if l passes through at least two of the existing facilities.

Med2: Since $\det(D) \neq 0$, we know that

$$Ex_m \in \mathcal{B}_l^- \Longleftrightarrow D(Ex_m) \in D(\mathcal{B}_l^-).$$

Recall that $\mathcal{B}_l^-$ and $\mathcal{B}_l^+$ are the two halfplanes separated by the line l (see Definition 8 on page 9). Since l is the boundary of both sets $\mathcal{B}_l^-$ and $\mathcal{B}_l^+$, we can use (3.5) to conclude that $D(\mathcal{B}_l^-) \in \{\mathcal{B}_{D(l)}^+, \mathcal{B}_{D(l)}^-\}$. Thus,two points Ex_m, Ex_k lie on the same side of l if and only if $D(Ex_m)$ and $D(Ex_k)$ lie on the same side of $D(l)$.

Cen1: An immediate consequence of (3.4) is that

$$d_t(Ex_k, l) \geq d_t(Ex_m, l) \Longleftrightarrow d_{e_2}(D(Ex_k), D(l)) \geq d_{e_2}(D(Ex_m), D(l)),$$

enforcing that l is at maximal (weighted) distance from at least three of the existing facilities if and only if $D(l)$ is at maximum (weighted) distance from at least three points in $D(\mathcal{E}x)$.

Cen2: From (3.5) it follows that f is a facet of some set A if and only if $D(f)$ is a facet of $D(A)$. Because of (3.4), $d(Ex_k, l) = d(Ex_m, l)$ if and only if $d(D(Ex_k), D(l)) = d(D(Ex_m), D(l))$ such that we can conclude:

l has the same slope as one of the facets (say, e.g., the facet defined by Ex_k and Ex_m) of the convex hull of the existing facilities if and only if $D(l)$ has the same slope as one of the facets (defined by $D(Ex_k)$ and $D(Ex_m)$) of the convex hull of the transformed existing facilities $D(\mathcal{E}x)$.

QED

For solving $1l/\mathbb{R}^2/\cdot/d_t/\sum$ we can proceed as follows: We choose a matrix D with $D(t) = e_2$ and $\det(D) \neq 0$. Then we define $Dx_m := D(Ex_m)$ for all $m \in \mathcal{M}$. With the new set of existing facilities $Dx_1, \ldots, Dx_m$ we solve the vertical line problem. Denoting the optimal solution by l^*, $D^{-1}(l^*)$ is an optimal solution of $1l/\mathbb{R}^2/\cdot/d_t/\sum$. The same procedure can also be applied to solve $1l/\mathbb{R}^2/\cdot/d_t/\max$.

In order to be able to extend the result of Theorem 3.1 to all distances derived from norms, we will prove that for any distance d derived from a norm γ and any line l there exists a direction $t \in \mathbb{R}^2$, depending only on the slope s of the line, such that

$$d(Ex_m, l) = d_t(Ex_m, l) \text{ for all } m \in \mathcal{M}.$$

This means, when evaluating the objective functions $f(l)$ or $g(l)$, we can replace d by d_t.

Lemma 3.10 *Let $s \in \mathbb{R} \cup \{\infty\}$ and let γ be a norm or $\gamma = \gamma_t$ for some vector $t \in \mathbb{R}^2$, such that t does not have the slope s. Let $d(x,y) = \gamma(y - x)$ be the corresponding distance. Then there exists a constant $C := C(s, d, l_2) \in \mathbb{R}, C > 0$, such that for all $x \in \mathbb{R}^2$ and all lines l with slope s*

$$d(x, l) = C\ l_2(x, l).$$

Proof: First consider the case that $x = 0$. Choose a line l^0 with slope s which does not contain the origin, i.e.,

$$l^0 = l_{s_1, s_2, b_0} \text{ with } -\frac{s_1}{s_2} = s \text{ and } b_0 \neq 0.$$

Then $l_2(0, l^0) \neq 0$ and $0 \neq d(0, l^0) < \infty$, and therefore we can find a real number $C > 0$ such that $d(0, l^0) = C l_2(0, l^0)$. Now take any other line l with slope s. Then there exists a $b \in \mathbb{R}$ such that $l = l_{s_1, s_2, b}$. Since $b_0 \neq 0$, there exists a real number β such that

$$b = \beta b_0.$$

Now we investigate the new line $\beta l_{s_1,s_2,b_0}$.

$$\begin{aligned}\beta l_{s_1,s_2,b_0} &= \{\beta x : x \in l_{s_1,s_2,b_0}\}\\ &= \{\beta(x_1,x_2) : s_1x_1 + s_2x_2 + b_0 = 0\}\\ &= \{(x_1',x_2') : \frac{s_1}{\beta}x_1' + \frac{s_2}{\beta}x_2' + b_0 = 0\}\\ &= \{(x_1',x_2') : s_1x_1' + s_2x_2' + \beta b_0 = 0\}\\ &= l_{s_1,s_2,\beta b_0} = l_{s_1,s_2,b}.\end{aligned}$$

Therefore we get the following equation.

$$\begin{aligned}|\beta| d(0, l_{s_1,s_2,b_0}) &= |\beta| \min_{y \in l_{s_1,s_2,b_0}} d(0,y)\\ &= \min_{y \in l_{s_1,s_2,b_0}} |\beta|\gamma(y)\\ &= \min_{y \in l_{s_1,s_2,b_0}} \gamma(|\beta| y)\\ &= \min_{y \in l_{s_1,s_2,b_0}} \gamma(\beta y) \qquad (3.6)\\ &= \min_{y' \in \beta l_{s_1,s_2,b_0}} \gamma(y')\\ &= \min_{y' \in l_{s_1,s_2,b}} \gamma(y')\\ &= d(0, l_{s_1,s_2,b})\end{aligned}$$

$$\Longrightarrow d(0, l_{s_1,s_2,b}) = |\beta| d(0, l_{s_1,s_2,b_0}) = |\beta| C l_2(0, l_{s_1,s_2,b_0}) = C l_2(0, l_{s_1,s_2,b}),$$

using the above equation for both d and l_2.
Since d is derived from a norm or from γ_t, it is translation invariant, and since the translation of a line does not affect its slope, we finally get $d(x, l_{s_1,s_2,b}) = Cl_2(x, l_{s_1,s_2,b})$ for any point $x \in \mathbb{R}^2$.

QED

Note that the properties of the Euclidean distance l_2 have not been used in the proof of in Lemma 3.10 and therefore l_2 can be replaced by any other distance derived from a norm or by distances derived from γ_t (if the slope of t differs from s). If d_1, d_2, and d_3 are distances with this property, we get that

$$C(s, d_1, d_2) = \frac{C(s, d_1, d_3)}{C(s, d_2, d_3)}.$$

Example 3.1 *Some examples demonstrate the calculation of* C.

- $C(s, d_{hor}, d_{ver}) = |s|$, *if* $s \in \mathbb{R} \setminus \{0\}$. $C(0, d_{hor}, d_{ver})$ *and* $C(\infty, d_{hor}, d_{ver})$ *are not defined since the slopes of* e_1 *and* e_2 *are* 0 *and* ∞ *(see the assumptions of Lemma 3.10.)*
- *For all* $s \in \mathbb{R}$

$$C(s, d_{ver}, l_p) = \sqrt[p]{\left(1 + |s|^{\frac{p}{p-1}}\right)^{p-1}}, \text{ see Lemma 3.5.}$$

- *$C(\infty, l_2, l_p) = 1$ for all $p \geq 1$, because $l_p(x, l_{s_1,0,b})$ is independent of p, see Lemma 3.5.*
- *$C(s, \gamma_1, \gamma_2)$ is defined for all $s \in \mathbb{R} \cup \{\infty\}$ if γ_1 and γ_2 are both norms.*

We remark that the directional bias function of a norm γ with derived distance function d is defined as

$$r_\gamma(x,y) = \frac{d(y-x)}{l_2(x,y)},$$

see [BL93]. In this sense,

$$r_\gamma(x,l) = \frac{d(x,l)}{l_2(x,l)} = C(l, l_2, \gamma)$$

can be interpreted as a directional bias function of the norm γ which in the case of lines is not dependent on x.

Theorem 3.2 *For a given slope $s \in \mathbb{R} \cup \{\infty\}$ all median lines with this slope are independent from the used norm. I.e., the lines l_{s_1,s_2,b^*} with $s = -\frac{s_1}{s_2}$ minimizing*

$$\min_{b \in \mathbb{R}} f(l_{s_1,s_2,b})$$

are identical for all distances d derived from norms. This result also holds for the center lines with fixed slope s.

Proof: The result follows directly from Lemma 3.10.

QED

The connection between any norm γ (with the derived metric $d(x,y) = \gamma(y-x)$) and the distances d_t becomes clear with the following lemma.

Lemma 3.11 *Let γ be a norm and d the corresponding distance. Let $x \in \mathbb{R}^2$ and l be a line. Then*

$$d(x,l) = \min_{t \in \mathbb{R}^2, \gamma(t)=1} d_t(x,l).$$

Proof: Let B be the unit ball of γ. From Lemma 3.7 we know that

$$d(x,l) = \min\{|\lambda| : x + \lambda B \cap l \neq \emptyset\} =: \lambda^0.$$

Thus there exists a $t^0 \in \mathbb{R}^2$ with $\gamma(t^0) = 1$ such that $x + \lambda^0 t^0 \in l$. (Note that $\gamma(t) = 1$ if and only if $t \in \partial(B)$.) Using the definition of d_{t^0}, this leads to

$$d(x,l) = d_{t^0}(x,l).$$

For all t' with $\gamma(t') = 1$ we calculate

$$\begin{aligned} d_{t'}(x,l) &= \min\{|\lambda| : x + \lambda t' \in l\} \\ &\geq \min\{|\lambda| : x + \lambda B \cap l \neq \emptyset\} \\ &= d(x,l), \end{aligned}$$

using Lemma 3.7 a second time.

QED

Lemma 3.12 *Let γ be a norm and d the corresponding distance. Let l be a line. Then there exists a $t \in \mathbb{R}^2$ with $\gamma(t) = 1$ such that*

$$d(x, l) = d_t(x, l) \text{ for all } x \in \mathbb{R}^2.$$

Proof: Let $x \in \mathbb{R}^2$ and $s \in \mathbb{R} \cup \{\infty\}$ be the slope of the line l. According to Lemma 3.11 we can find a direction $u \in \mathbb{R}^2$ such that $\gamma(u) = 1$ and

$$d(x, l) = d_u(x, l) \leq d_t(x, l) \text{ for all t} \in \mathbb{R}^2 \text{ with } \gamma(t) = 1.$$

Now suppose there exist $y \in \mathbb{R}^2$ and $v \in \mathbb{R}^2$ with $\gamma(v) = 1$ and

$$d(y, l) = d_v(y, l) < d_u(y, l).$$

Note that u and v both do not have slope s, because $d(x, l)$ and $d(y, l)$ both are not infinite. With Lemma 3.10 we can conclude that there exists a constant $C := C(s, d_u, d_v)$ such that

$$\begin{aligned} d_u(y, l) &= C d_v(y, l) \text{ and} \\ d_u(x, l) &= C d_v(x, l), \end{aligned}$$

yielding $C > 1$ in the first case and $C \leq 1$ in the second case, a contradiction.

QED

Note that, if there exists an $x \in \mathbb{R}^2$ with $d(x, l) = d_t(x, l)$ for a direction $t \in \mathbb{R}^2$, then the proof of Lemma 3.12 shows that $d(y, l) = d_t(y, l)$ for all $y \in \mathbb{R}^2$. Now we are ready to prove the main result of this section, namely that the properties Med1, Med2, Cen1, and Cen2 hold for all distances derived from norms.

Theorem 3.3 *For all distances d derived from norms the following criteria hold.*

Med1: *There exists a median line passing through two of the existing facilities.*

Med2: *All median lines are pseudo-halving.*

Cen1: *There exists a center line which is at maximum distance from three of the existing facilities.*

Cen2: *If the weights are all equal, there exists a center line which is parallel to a facet of $conv(\mathcal{E}x)$.*

Proof:

Med1: Suppose l^* is an optimal line, but does not pass through two of the existing facilities. Choose t^* such that $d(Ex_m, l^*) = d_{t^*}(Ex_m, l^*)$ for all $m \in \mathcal{M}$ according to Lemma 3.12.

From Theorem 3.1 we know that Med1 holds for $1l/\mathbb{R}^2/\cdot/d_{t^*}/\sum$, and therefore we can choose l^0 minimizing the distance d_{t^*} and passing through at least two of the existing facilities.

Now let t^0 be such that $d(Ex_m, l^0) = d_{t^0}(Ex_m, l^0)$ for all $m \in \mathcal{M}$ according to Lemma 3.12. Then we get:

$$\begin{aligned} f(l^*) &= \sum_{m\in\mathcal{M}} w_m d(Ex_m, l^*) \\ &= \sum_{m\in\mathcal{M}} w_m d_{t^*}(Ex_m, l^*) \\ &\geq \sum_{m\in\mathcal{M}} w_m d_{t^*}(Ex_m, l^0) \\ &\geq \sum_{m\in\mathcal{M}} w_m d_{t^0}(Ex_m, l^0) \text{ because of Lemma 3.11} \\ &= \sum_{m\in\mathcal{M}} w_m d(Ex_m, l^0) \\ &= f(l^0) \geq f(l^*) \text{ because of the optimality of } l^*. \end{aligned}$$

Thus l^0 is an optimal line, too, that passes through two of the existing facilities, and the proof of Med1 is complete.

Cen1 and Cen2: The proof of Cen1 and Cen2 is analogous to the proof of Med1. Replacing $\sum$ by max in the above formulas and using the property *to be at maximal (weighted) distance from three of the existing facilities* for Cen1, and *to pass through one facet of the convex hull of the existing facilities* for Cen2 instead of *passing through two of the existing facilities* in the above proof, we analogously construct a line l^0 with $g(l^*) \geq g(l^0) \geq g(l^*)$.

Med2: We assume that there exists a median line l^* with $\sum_{Ex_m \in \mathcal{B}^-_{l^*}} w_m > \frac{W}{2}$. With the same notation as in the proof of Med1 we can conclude from Theorem 3.1 that any line l^0 minimizing d_{t^*} satisfies

$$\sum_{Ex_m \in \mathcal{B}^-_{l^0}} w_m \leq \frac{W}{2} \text{ and } \sum_{Ex_m \in \mathcal{B}^+_{l^0}} w_m \leq \frac{W}{2}.$$

Since l^* cannot be optimal for $1l/\mathbb{R}^2/\cdot/d_{t^*}/\sum$ according to Theorem 3.1, we get

$$f(l^*) = \sum_{m\in\mathcal{M}} w_m d_{t^*}(Ex_m, l^*)$$

$$
\begin{aligned}
&> \sum_{m \in \mathcal{M}} w_m d_{t^*}(Ex_m, l^0) \\
&\geq f(l^0),
\end{aligned}
$$

contradicting the optimality of l^*.

QED

3.3 Algorithmic Approaches for Norms

Lemma 3.7 shows that the distance $d(x, l)$ between a point x and a hyperplane l depends on the shape of the unit ball B which can be an arbitrary convex set. Thus, for certain unit balls (e.g., unit balls having smooth boundary whose analytical description might be sufficiently complicated) the calculation of $d(x, l)$ is impossible by using discrete methods in the spirit of computational geometry. On the other hand, there are norms (like the Euclidean one) that give direct motivation and basis for computational approaches, and we will show in Section 3.4 that for block norms the time complexity of solution algorithms is even more reducible. For the following computational arguments we will ignore these calculation difficulties. And so it turns out that the results of Theorem 3.3 yield the possibility to solve line location problems with median and with center objective function by enumerating a finite set of candidates. Although such an enumeration approach may not be very efficient for some special norms (e.g., for the l_1 norm much faster algorithms exist), it has the advantage that it can be used for all distances derived from norms.

Algorithm 1 (for solving $1l/\mathbb{R}^2/\cdot/\gamma/\sum$)

Input: Norm γ, $Ex_m, w_m > 0$ for all $m \in \mathcal{M}$.

Output: A line l^* with objective value z^* which solves $1l/\mathbb{R}^2/\cdot/\gamma/\sum$

1. $z^* := \infty$.

2. For all lines l passing through two of the existing facilities do

1. Determine $f(l)$.
2. If $f(l) < z^*$ then $z^* := f(l)$ and $l^* := l$.

3. Output: l^* with objective value z^*.

For the center problem with arbitrary weights we propose Algorithm 2. To this end, we first introduce

$$
P_{jk} := \frac{w_j}{w_j + w_k} Ex_j + \frac{w_k}{w_j + w_k} Ex_k
$$

for all $j, k \in \mathcal{M}, j \neq k$. Note that $P_{jk} = P_{kj}$.

Lemma 3.13 *Let l be a line and $j, k \in \mathcal{M}$. For all distances d derived from norms the following relations hold:*

1. $P_{jk} \in l \Longrightarrow w_j d(Ex_j, l) = w_k d(Ex_k, l)$.
2. *If $d(Ex_j, l) = w_k d(Ex_k, l)$ and l separates Ex_j from Ex_k then $P_{jk} \in l$.*

Proof:

1. Let $P_{jk} \in l$. If both Ex_k and Ex_j are contained in l the result is trivial. Now we assume that l separates Ex_k and Ex_l and hence that the slope of $Ex_k - Ex_j$ is not the same as the slope s of the line l. Define

$$t = \frac{Ex_k - Ex_j}{\|Ex_k - Ex_j\|}.$$

We then conclude that

$$\begin{aligned} d_t(Ex_j, l) &= d_t(Ex_j, P_{jk}) = l_2(Ex_j, P_{jk}) \text{ and} \\ d_t(Ex_k, l) &= d_t(Ex_k, P_{jk}) = l_2(Ex_k, P_{jk}), \end{aligned}$$

according to Formula 3.3 on page 59. Furthermore, from the definition of P_{kl} it follows that $w_j l_2(Ex_j, P_{jk}) = w_k l_2(Ex_k, P_{jk})$ and hence

$$w_j d_t(Ex_j, l) = w_k d_t(Ex_k, l).$$

Now consider any distance d derived from a norm γ. Since s is unequal to the slope of t we can apply Lemma 3.10 (on page 62) and choose $C := C(s, \gamma, \gamma_t)$ such that $d(x, l) = C d_t(x, l)$ for all $x \in \mathbb{R}^2$. Summarizing we get

$$\begin{aligned} w_j d(Ex_j, l) &= w_j C d_t(Ex_j, l) \\ &= w_k C d_t(Ex_k, l) \\ &= w_k d(Ex_k, l). \end{aligned}$$

2. Let $w_j d^0(Ex_j, l) = w_k d^0(Ex_k, l)$ for a distance d^0 derived from the norm γ^0. As in the first part of this proof, define $t = \frac{Ex_k - Ex_j}{\|Ex_k - Ex_j\|}$. If s denotes the slope of l there exists a constant $C := C(s, \gamma_t, \gamma^0)$ according to Lemma 3.10 such that

$$d_t(x, l) = C d^0(x, l) \text{ for all } x \in \mathbb{R}^2.$$

(Note that the slope of t is not equal to s since l separates Ex_j and Ex_k.) Hence we also have

$$w_j d_t(Ex_j, l) = w_k d_t(Ex_k, l).$$

Since l separates Ex_k from Ex_j we conclude that l intersects the line segment from Ex_k to Ex_j in a unique point P. Hence we can calculate that

$$\begin{aligned} d_t(Ex_j, l) &= d_t(Ex_j, P) = l_2(Ex_j, P) \text{ and} \\ d_t(Ex_k, l) &= d_t(Ex_k, P) = l_2(Ex_k, P). \end{aligned}$$

I.e., $w_j l_2(Ex_j, P) = w_k l_2(Ex_k, P)$ yielding that $P = P_{jk}$.

QED

Corollary 3.1 *For all distances d derived from norms the following holds: A line l is at maximum (weighted) distance from the existing facilities Ex_j, Ex_k and Ex_l while separating Ex_j from Ex_k and Ex_l if l passes through P_{jk} and through P_{jl}.*

Theorem 3.4 *For all distances d derived from norms the following holds: There exists a center line l and $j, k, l \in \mathcal{M}$ such that l passes through P_{jk} and through P_{jl}.*

Proof: First we note that Theorem 3.4 is correct in the trivial case that all existing facilities lie on the same (optimal) line with objective value $g(l) = 0$. If $g(l) > 0$ we know from Theorem 3.3 that there exists a center line l which is at maximum distance from three of the existing facilities, e.g. from Ex_j, Ex_k, and Ex_l. Obviously, not all three points Ex_j, Ex_k, and Ex_l can lie on the same side of l, because the objective value $g(l)$ then could be improved by moving l parallel towards Ex_j, Ex_k, and Ex_l. Hence we can assume that l separates Ex_j from Ex_k and from Ex_l and thus apply Corollary 3.1. This yields that $P_{jk}, P_{jl} \in l$.

QED

Thus we have established the validity of the following algorithm for solving center problems.

Algorithm 2 (for solving $1l/\mathbb{R}^2/\cdot/\gamma/\max$)

Input: Norm γ, $Ex_m, w_m > 0$ for all $m \in \mathcal{M}$.

Output: A line l^* with objective value z^* which solves $1l/\mathbb{R}^2/\cdot/\gamma/\max$

1. $z^* := \infty$.

2. Determine $\mathcal{P} = \{P_{jk} : j, k \in \mathcal{M}, j \neq k\}$.

3. For all lines l passing through two points $P_{jk}, P_{jl} \in \mathcal{P}$, $P_{jk} \neq P_{jl}$ do

1. Calculate $g(l)$.
2. If $g(l) < z^*$ then $z^* := g(l)$ and $l^* := l$.

4. Output: l^* with objective value z^*.

In Step 3 we assumed $P_{jk} \neq P_{jl}$. If such two points coincide, the line passing through $P_{jk} = P_{jl}$ is not uniquely defined. In this case, Ex_j, Ex_k, and Ex_l are collinear. Now take any line l which is at maximum distance from Ex_j, Ex_k, and Ex_l and assume $l \cap \mathcal{P} = \{P_{jk}, P_{jl}\}$, i.e.,

$$\begin{aligned} w_m d(Ex_m, l) &= g(l) \text{ for } m \in \{j,k,l\} \text{ and} \\ w_m d(Ex_m, l) &< g(l) \text{ for } m \notin \{j,k,l\}. \end{aligned}$$

Then l cannot be optimal, since by rotating l about P_{jk} it is possible to improve its distance to all three existing facilities Ex_j, Ex_k, and Ex_l and hence its objective value. Note that in Chapter 7 we will prove that there always exists a line which has maximum distance to 3 *affinely independent* existing facilities.

The efficiency of Algorithms 1 and 2 can be improved. By enumerating only the pseudo-halving lines in Algorithm 1, similar to the Euclidean procedure described on page 49, one gets a running time of $O(M^2 \log M)$. For the center problem it is sufficient to investigate those points P_{jk} in Algorithm 2 that may really lead to a maximum distance. For the Euclidean case [LW86] characterizes such a subset of $\mathcal{P}$ by giving a reduction from the planar weighted Euclidean problem to the unweighted case. Finally, in the special case of equal weights, the center problem can be solved by Algorithm 3, the validity of which is based on the parallel facet property Cen2.

Algorithm 3 (for solving $1l/\mathbb{R}^2/w_m = 1/\gamma/\max$)

Input: Norm γ with derived distance d, $Ex_m, w_m > 0$ for all $m \in \mathcal{M}$.

Output: A line l^* with objective value z^* which solves $1l/\mathbb{R}^2/w_m = 1/\gamma/\max$

1. Determine the convex hull of the set of the existing facilities together with its extreme points $Ext(conv(\mathcal{E}x))$ and its edges.

2. $z^* := \infty$.

3. For all antipodal pairs (v, e) of $conv(\mathcal{E}x)$ with $e = (v_1, v_2)$ and with $v, v_1, v_2 \in Ext(conv(\mathcal{E}x))$ do

1. Determine the line l passing through v_1 and v_2.
2. Calculate the distance $d(v, l)$.
3. If $d(v,l) < z^*$ then $z^* := d(v,l)$ and $l^* := l$, shifted by $\frac{z^*}{2}$ towards v.

4. Output: l^* with objective value $z^* := \frac{z^*}{2}$.

By applying the method of rotating callipers originally developed in [Sha78] and also used in [Tou83] in order to solve set width problems the time complexity of

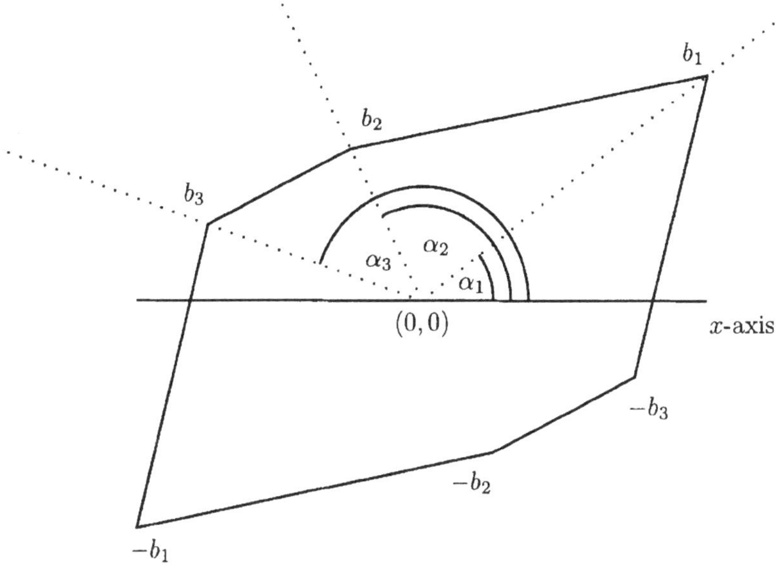

Figure 3.2. A block norm with $G = 3$.

Algorithm 3 can be reduced to $O(MlogM)$ which has been shown to be optimal in the Euclidean case (see [LW86]).

3.4 A Fast Algorithm for Block Norm Distances

In the special case that the distance measure d is derived from a block norm (i.e., the unit ball B is a polytope, see Definition 3 on page 4) it is possible to find median and center lines very efficiently, although the corresponding line location problems are not convex. But it is possible to use the convexity of the corresponding line location problems with vertical distances d_{ver} to get an efficient approach.

Lemma 3.14 *Let d_B be derived from a block norm γ_B with fundamental directions $Ext(B) := \{b_1, b_2, \ldots, b_G\}$ and let l be a line. Then there exists a $g \in \{1, 2, \ldots, G\}$ such that for all $x \in \mathbb{R}^2$*

$$d_B(x, l) = d_{b_g}(x, l).$$

Proof: In order to determine $d(x, l)$ we know from Lemma 3.7 that the unit ball can be dilated around x until it touches l. As d_B is derived from a block norm, the unit ball B of d_B is a polytope. Obviously, a line touches a polytope in at least one vertex of that polytope, see, e.g., [Sha78]. Thus the formula

$$d(x,l) = \min_{t \in \mathbb{R}^2, \gamma(t)=1} d_t(x,l)$$

from Lemma 3.11 simplifies in the case of block norms to

$$d_B(x,l) = \min_{g=1,\ldots,G} d_{b_g}(x,l).$$

With Lemma 3.12 we conclude that there exists a $b_g \in Ext(B)$ such that

$$d_B(x,l) = d_{b_g}(x,l) \text{ for all } x \in \mathbb{R}^2.$$

QED

With the help of the above lemma we can decompose our problem in G independent subproblems. Thus, to solve $1l/\mathbb{R}^2/\cdot/d_B/\sum$ and $1l/\mathbb{R}^2/\cdot/d_B/\max$, respectively, we can proceed as follows. For all $g = 1,2,\ldots,G$ we solve the corresponding location problem with distance d_{b_g}, i.e. we find the best line l_g^* minimizing

in the median case: $\sum_{m\in\mathcal{M}} w_m d_{b_g}(Ex_m, l)$ and

in the center case: $\max_{m\in\mathcal{M}} w_m d_{b_g}(Ex_m, l)$.

Then we choose l^* as the line l_g^*, $g = 1,2,\ldots,G$ with the smallest objective value.

In order to find an optimal line l_g^* of $1l/\mathbb{R}^2/\cdot/d_{b_g}/\sum$ and $1l/\mathbb{R}^2/\cdot/d_{b_g}/\max$, respectively, we can use Lemma 3.9 of the previous section (see page 60). More particular, let us choose the transformation D as a rotation and stretching according to Lemma 3.15 below. Recall that $\|t\|$ denotes the Euclidean length of a vector $t \in \mathbb{R}^2$. Furthermore, let α_g be defined as the angle between b_g and the x–axis, for all $g = 1,2,\ldots,G$. For an illustration of α_g, see Figure 3.2. Now we are ready to define the following transformation.

Lemma 3.15 *Define the following rotation D^r and the rotation stretching D^{rs}:*

$$D^r = \begin{pmatrix} sin(\alpha_g) & -cos(\alpha_g) \\ cos(\alpha_g) & sin(\alpha_g) \end{pmatrix}$$

and $D^{rs} = \frac{1}{|b_g|} D^r$. Then

$$d_{b_g}(x,l) = d_{ver}(D^{rs}(x), D^{rs}(l)) = \frac{1}{|b_g|} d_{ver}(D^r(x), D^r(l)).$$

Proof: According to Lemma 3.9 we have to verify that $D^{rs}(b_g) = e_2$ and $det(D^{rs}) \neq 0$. This holds since D^{rs} rotates all vectors about the origin with the angle $90° - \alpha$ in mathematically positive direction and stretches them by the factor $\frac{1}{|b_g|}$. Furthermore $det(D^{rs}) = \frac{1}{|b_g|} \neq 0$. QED

Therefore we get the following algorithm.

Algorithm 4 (for solving $1l/\mathbb{R}^2/\cdot/\gamma_B/\sum$ and $1l/\mathbb{R}^2/\cdot/\gamma_B/\max$)

Input: Block norm distance d_B, $Ex_m, w_m > 0$ for all $m \in \mathcal{M}$

Output: A line l^* with objective value z^* which

for the median problem : solves $1l/\mathbb{R}^2/\cdot/d_B/\sum$,

for the center problem : solves $1l/\mathbb{R}^2/\cdot/d_B/\max$.

1. $z^* := \infty$

2. For $g = 1$ to G do

1. Calculate the angle α_g between b_g and the x–axis.
2. $D = \begin{pmatrix} sin(\alpha_g) & -cos(\alpha_g) \\ cos(\alpha_g) & sin(\alpha_g) \end{pmatrix}$
3. For $m \in \mathcal{M}$ do: $Dx_m = D(x_m)$.
4. Find a line l_g^* minimizing

   for the median problem : $f(l) = \sum_{m \in \mathcal{M}} w_m d_{ver}(Dx_m, l)$,

   for the center problem : $f(l) = \max_{m \in \mathcal{M}} w_m d_{ver}(Dx_m, l)$.

   (Use the algorithms mentioned in Section 2.1.)
5. If $f(l_g^*) < z^*|b_g|$ then set

$$z^* := \frac{f(l_g^*)}{|b_g|}$$

   and $l^* := D^{-1}(l_g^*)$.

3. Output: l^* with objective value z^*

The correctness of this algorithm follows from Lemma 3.14. We remark that Med1, Med2, Cen1, and Cen2 can be directly derived from this algorithm for block norms, without using the more general Theorem 3.3. This was done by the author in a previous approach, see [Sch96].

Algorithm 4 runs in $O(GR)$, where R is the complexity to solve the corresponding problems with vertical distance ($1l/\mathbb{R}^2/\cdot/d_{ver}/\sum$ or $1l/\mathbb{R}^2/\cdot/d_{ver}/\max$, respectively). In [Zem84] it was shown that for the median problem this can be done in linear time, such that the above algorithm solves median problems with block norm distances in $O(GM)$ time.

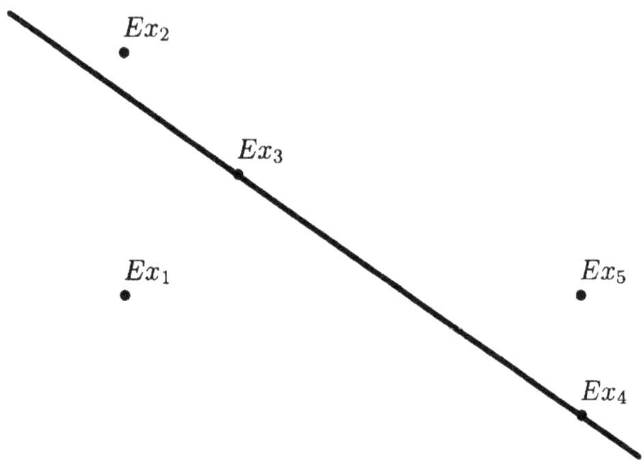

Figure 3.3. Optimal line in Example 3.2.

The center problem with vertical distance can also be solved in linear time using linear programming (cf. [Meg84]), and therefore the above Algorithm 4 solves $1l/\mathbb{R}^2/\cdot/d_B/\max$ in $O(GM)$ time, too.

A common and simple example for a block norm is l_1. The median problem denoted $1l/\mathbb{R}^2/\cdot/l_1/\sum$, i.e., to locate a line in the plane with rectangular distance, has been well solved by many authors, compare Section 3.1. We already mentioned in Section 3.1 that most algorithms for this problem separate it into two subproblems; namely to find the best lines minimizing the sum of vertical and the sum of horizontal distances, respectively, and then choosing the better one. Most of the l_1–algorithms in the plane also give a transformation from the horizontal line problem to the vertical line problem or vice versa. We remark that in this sense the usual algorithms for l_1 are special cases of Algorithm 4. In Section 7.3 we will show that this algorithm can be applied to solve line transversal problems efficiently.

To demonstrate Algorithm 4 we conclude this section with an example.

Example 3.2 *Let $\gamma_B = l_\infty$ be the Chebychev distance l_∞, i.e.,*

$$Ext(B) = \{(1,1), (-1,1), (-1,-1), (1,-1)\}.$$

Let $\mathcal{M} = \{1, 2, \ldots, 5\}$ and

m	1	2	3	4	5
Ex_m	$(1,1)$	$(1,3)$	$(2,2)$	$(5,0)$	$(5,1)$

,

(see Figure 3.3). Furthermore $w_m = 1$ *for all* $m \in \mathcal{M}$. *Now we determine a median line for this problem, i.e., we solve* $1l/\mathbb{R}^2/w_m = 1/l_\infty/\sum$. *We note that* $G = 2$, $b_1 = (1,1)$, *and* $b_2 = (-1,1)$. *Applying Algorithm 4 we get:*

For $g = 1 \; : \alpha_1 = 45°, \; D = \frac{1}{2}\sqrt{2}\begin{pmatrix} 1 & -1 \\ 1 & 1 \end{pmatrix}$ *and*

m	1	2	3	4	5
Dx_m	$(0,\sqrt{2})$	$(-\sqrt{2},2\sqrt{2})$	$(0,2\sqrt{2})$	$(\frac{5}{2}\sqrt{2},\frac{5}{2}\sqrt{2})$	$(2\sqrt{2},3\sqrt{2})$

Solving $1l/\mathbb{R}^2/\cdot/d_v/\sum$ *with these data we obtain*

$$l_1^* : x_2 = \frac{1}{5}x_1 + 2\sqrt{2} \text{ and } f(l_1^*) = \frac{9}{5}\sqrt{2}.$$

For $g = 2 \; : \alpha_2 = 135°, \; D = \frac{1}{2}\sqrt{2}\begin{pmatrix} 1 & 1 \\ -1 & 1 \end{pmatrix}$ *and*

m	1	2	3	4	5
Dx_m	$(\sqrt{2},0)$	$(2\sqrt{2},\sqrt{2})$	$(2\sqrt{2},0)$	$(\frac{5}{2}\sqrt{2},\frac{5}{2}\sqrt{2})$	$(3\sqrt{2},-2\sqrt{2})$

which leads to

$$l_2^* : x_2 = -x_1 + \sqrt{2} \text{ and } f(l_2^*) = 4\sqrt{2}.$$

Since

$$\frac{f(l_1^*)}{l_2(0,b_1)} < \frac{f(l_2^*)}{l_2(0,b_2)}$$

we get

$$\text{Output: } z^* = \frac{f(l_1^*)}{l_2(0,b_1)} = \frac{9}{5} \text{ and } l^* = D^{-1}(l_1^*) : x_2 = -\frac{2}{3}x_1 + \frac{10}{3}.$$

This line passes through the existing facilities Ex_3 *and* Ex_4.

4 FINDING ALL OPTIMAL LINES

4.1 The Strong Incidence Property

The main result for the median problem — the weak incidence property or Med1 — has been discussed in Section 3.2. We have shown that for all distances derived from norms and all sets of existing facilities $\mathcal{E}x$ there *exists* a line minimizing the sum of distances to all points in $\mathcal{E}x$ and passing through two of the existing facilities.

For d being the Euclidean distance a much stronger criterion holds. Korneenko and Martini showed that for l_2 *all* median lines pass through two of the existing facilities, see [KM90] and [KM93]. For their proof they used the theory of zonotopes, i.e., vector sums of line segments or, equivalently, convex n-polytopes whose r-faces, $2 \leq r \leq n-1$, are all centrally symmetric. In [Mar87] Martini observed a relationship between support functions of zonotopes and the Euclidean line location problem. Then, using necessary conditions for local minima of these support functions, Korneenko and Martini proved that *every* optimal $(n-1)$-subspace passes through $n-1$ linearly independent existing facilities. Identifying one given point as the "new" origin yields the following necessary condition for the solutions of $1H/\mathbb{R}^n/\cdot/l_2/\sum$, due to [KM90] (see also [KM93]): In Euclidean n-space, *every* optimal hyperplane has to pass through n affinely independent points of the given set. For $n = 2$ this is equivalent to the above mentioned result — the strong incidence property — for locating a median line with Euclidean distance. Note that Lemma 3.6 shows that the strong incidence property holds also in the planar case of p-norm distances with $p \in (0, 1)$.

This strong incidence property cannot be generalized to all distances derived from norms. An easy example is the following.

Example 4.1 *Consider Example 1.2 on page 16 and Figure 1.2, i.e., an unweighted problem with rectangular distance in the plane* $(1l/\mathbb{R}^2/w_m = 1/l_1/\sum)$, *where* $\mathcal{E}x$ *is given by the four vertices of a square* Q, *i.e.,*

$$Ex_1 = (-1, 1), Ex_2 = (1, -1), Ex_3 = (1, 1), \text{ and } Ex_4 = (-1, 1).$$

All lines l_{ij} *passing through two existing facilities* Ex_i *and* Ex_j *have the same objective value* $f(l_{ij}) = \sum_{m=1}^{4} l_1(Ex_m, l_{ij}) = 4$. *Therefore 4 is the optimal value and all six lines* $l_{ij}, i < j$, *are optimal. But also all lines* l *intersecting either the two vertical facets or the two horizontal facets of the square* Q *have the same sum of distances to the four given existing facilities and are thus also optimal. Note that in this example all pseudo-halving lines are optimal solutions.*

Hence there exist distances derived from norms where we can construct median lines which contain *no* point of a suitably chosen set of existing facilities. A natural question that arises in this context is to ask for geometric characterizations of those norms (or unit balls) which enforce the stronger incidence property, that *each* median line necessarily passes through two of the existing facilities. This question will be answered in the following.

Definition 19 (e.g., [Phe89]) *Let* γ *be a norm and let* $B = \{x : \gamma(x) \leq 1\}$ *be the unit ball of* γ. γ *is called a* **smooth norm** *if for all* $t \in \mathbb{R}^2$ *with* $\gamma(t) = 1$ *there exists exactly one supporting line of* B *at* t, *i.e.,* B *is a* **smooth set**.

Examples for smooth norms are the l_p norms with $1 < p < \infty$, whereas no block norm is a smooth norm. We can use the following lemma to find supporting lines of B. Recall that d_t is defined in Definition 18 on page 58 and the t-distance between a point and a line can be calculated according to Lemma 3.8.

Lemma 4.1 *Let* d *be a distance derived from a norm* γ. *Let* $t, x \in \mathbb{R}^2$ *be such that* $\gamma(t) = 1$ *and* $d(x, l) = d_t(x, l)$. *Then there exists a translate of* l *which is a supporting line of the unit ball* B *of* γ *at* t.

Proof: Since we investigate translates of l we can assume $x = 0$. Let $d(0, l) = \lambda$. This means that the ball λB around x touches the line l (Lemma 3.7). Since $d_t(0, l) = \lambda$ we know from Lemma 3.8 that $\lambda t \in l$. We also have

$$\gamma(t) = 1 \Longrightarrow \lambda t \in \partial(\lambda B).$$

Summarizing we can conclude that l is a supporting line of λB at λt and therefore a translate of l supports B at t.

QED

In the following we will show that the strong incidence property holds exactly for smooth norms. For this purpose we need one more lemma for the problem with vertical(!) distance d_{ver}.

Lemma 4.2 *If* $1l/\mathbb{R}^2/\cdot/d_{ver}/\sum$ *has two distinct optimal lines then it also has two non-parallel optimal lines.*

Proof: We use the dual interpretation of our problem that has been introduced in Section 2.1. According to Theorem 1.3 (on page 24) we know that the optimal solution is either a vertex of a cell in the dual space — in this case it is unique — or it is a complete edge or a complete cell. Since parallel lines have the same slope they are represented by points on the same vertical line in the dual space. As none of the construction lines $T(Ex_m) = l_{-a_{m1},-1,a_{m2}}$ is a vertical line (all first coordinates of the existing facilities are real numbers) there are no cells with vertical edges. If the optimal solution is not unique, there exist two optimal solutions in the dual space with different first coordinates, representing two lines l^1 and l^2 which are both optimal and have different slopes.

QED

Corollary 4.1 *Let* $t \in \mathbb{R}^2 \setminus \{0\}$. *If* $1l/\mathbb{R}^2/\cdot/d_t/\sum$ *has two distinct optimal lines then it also has two non-parallel optimal lines.*

Proof: This follows from Lemma 4.2 and by using a transformation D as described in Lemma 3.9.

QED

Now we can prove the main result of this section.

Theorem 4.1 *γ is a smooth norm if and only if the strong incidence property is true for all instances of* $1l/\mathbb{R}^2/\cdot/\gamma/\sum$. *I.e., if l^* is a median line, then l^* passes through two of the existing facilities.*

Proof:

$\Longrightarrow$: Let γ be a smooth norm and $d(x,y) = \gamma(y-x)$ be the derived distance. Let $\mathcal{E}x = \{Ex_1, Ex_2, \ldots, Ex_M\}$ be a set of existing facilities with weights $w_m \geq 0$ for $m = 1,2,\ldots,M$. Now we consider an optimal line l^* for $1l/\mathbb{R}^2/\cdot/\gamma/\sum$. Let s^* be the slope of l^*.

Suppose that l^* does not contain two of the existing facilities. We choose $t^* \in \mathbb{R}^2$ with $\gamma(t^*) = 1$ according to Lemma 3.12 such that

$$d_{t^*}(Ex_m, l^*) = d(Ex_m, l^*) \text{ for all } m \in \mathcal{M}.$$

Now we consider the same instance of the location problem, but with distance d_{t^*} instead of d ($1l/\mathbb{R}^2/\cdot/d_{t^*}/\sum$). For this problem the weak incidence property holds (see Theorem 3.1), such that we can choose a line

l^0 that is also optimal for $1l/\mathbb{R}^2/\cdot/\gamma/\sum$ and that passes through two of the existing facilities, thus $l^* \neq l^0$. Since l^* and l^0 are both median lines, Corollary 4.1 implies that there exists a line l' with slope s' such that l' is also optimal for the problem instance with distance d_{t^*} and $s^* \neq s'$. Now choose t' with $\gamma(t') = 1$ such that

$$d_{t'}(Ex_m, l') = d(Ex_m, l') \text{ for all } m \in \mathcal{M},$$

again using Lemma 3.12. Then, according to Lemma 3.11,

$$d_{t^*}(Ex_m, l') \geq d_{t'}(Ex_m, l') \text{ for all } m \in \mathcal{M}.$$

Furthermore, since γ is a smooth norm,

$$d_{t^*}(Ex_m, l') \neq d_{t'}(Ex_m, l') \text{ for all } m \in \mathcal{M},$$

because otherwise $d(Ex_m, l') = d_{t'}(Ex_m, l') = d_{t^*}(Ex_m, l')$, meaning that there would exist two different supporting lines of B at t^* according to Lemma 4.1, namely:

- a translate of l^*, as $d(Ex_m, l^*) = d_{t^*}(Ex_m, l^*)$ and
- a translate of l', as $d(Ex_m, l') = d_{t^*}(Ex_m, l')$.

Summarizing we get $d_{t^*}(Ex_m, l') > d_{t'}(Ex_m, l')$ for all $m \in \mathcal{M}$, and therefore

$$\begin{aligned} f(l^*) &= \sum_{m\in\mathcal{M}} w_m d(Ex_m, l^*) \\ &= \sum_{m\in\mathcal{M}} w_m d_{t^*}(Ex_m, l^*) \\ &= \sum_{m\in\mathcal{M}} w_m d_{t^*}(Ex_m, l') \\ &> \sum_{m\in\mathcal{M}} w_m d_{t'}(Ex_m, l') \\ &= \sum_{m\in\mathcal{M}} w_m d(Ex_m, l') = f(l'), \end{aligned}$$

contradicting the optimality of l^*.

$\Longleftarrow$: Now let γ be a non-smooth norm. Then there exists a $t \in \mathbb{R}^2$ with $\gamma(t) = 1$ such that there are two different supporting lines of B at t. Let $\mathcal{J} = [\underline{s}, \overline{s}]$ be the interval of all slopes of supporting lines of B at t. Without loss of generality let $0 \in int(\mathcal{J})$. (This can be achieved by rotation of the coordinate system.) Now consider the following four existing facilities (see also Figure 4.1).

$$Ex_1 = (0,0), Ex_2 = t, Ex_3 = (c,0) + t \text{ and } Ex_4 = (c,0).$$

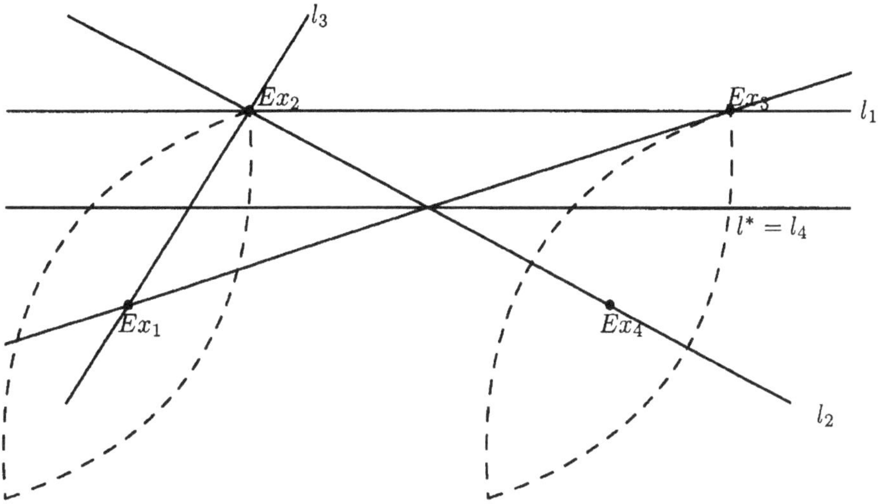

Figure 4.1. For non-smooth norms there can exist median lines which do not contain any of the existing facilities.

Let c be sufficiently large so that translates of both diagonals of the parallelogram (the lines through Ex_2 and Ex_4 and through Ex_1 and Ex_3, respectively) are supporting lines of B at t. This is possible since $0 \in int(\mathcal{J})$. Now we determine the objective values for all lines passing through two of the existing facilities. Because of the symmetry it is sufficient to consider

$$\begin{aligned} l_1 &= \text{the line through } Ex_2 \text{ and } Ex_3 \\ l_2 &= \text{the line through } Ex_2 \text{ and } Ex_4 \\ l_3 &= \text{the line through } Ex_2 \text{ and } Ex_1. \end{aligned}$$

It holds that $d(Ex_1, l_1) = \gamma(t) = d(Ex_1, l_2)$, because l_1 and l_2 are supporting lines of B at t, and, furthermore, $d(Ex_4, l_1) = d(Ex_1, l_1) = \gamma(t)$. Since γ is symmetric with respect to the origin we have $d(Ex_3, l_2) = d(Ex_1, l_2) = \gamma(t)$. If c is sufficiently large we therefore get

$$f(l_3) > 2\gamma(t) = f(l_1) = f(l_2),$$

yielding that both l_1 and l_2 are optimal lines. Now consider

$$l_4 \quad = \quad \text{the line through } \frac{1}{2}t \text{ and } Ex_4 + \frac{1}{2}t.$$

We calculate that

$$d(Ex_m, l_4) = \gamma(\frac{t}{2}) = \frac{1}{2}\gamma(t) \text{ for } m = 1, 2, 3, 4,$$

and consequently

$$f(l_4) = 4 \cdot \frac{1}{2}\gamma(t) = 2\gamma(t) = f(l_1) = f(l_2).$$

It follows that the line l_4 is optimal, too. But l_4 does not contain any of the existing facilities, yielding that the strong incidence property does not hold.

QED

A similar proof for the center problem $1l/\mathbb{R}^2/\cdot/\gamma/\max$ and the strong blockedness property is possible. While the proof of Theorem 4.1 can be applied in exactly the same way for the center objective, Lemma 4.2 is not so easy to prove with the help of the dual interpretation. Since we need it for the location of hyperplanes, this lemma will be proved in the higher dimensional case (Lemma 7.5) in Chapter 7. Nevertheless, we will derive the result for the center objective in the next section, but by using purely geometric arguments giving further insight into the center problem. Furthermore, these geometric tools will also be needed in Chapter 5, where we cannot use the same argumentation as for the median case. (In particular, we will show that there exist distance measures for which Med1 does not hold, but Cen1 does.)

Finally, we will transfer the property that all optimal lines are halving lines (as defined in Definition 9 on page 15) from the Euclidean distance to all distances which are derived from smooth norms. (For distances that are derived from non-smooth norms, this statement is not true, see Example 4.1.)

Corollary 4.2 *Consider the median problem with distance d derived from a smooth norm γ. Then all optimal lines are halving lines.*

Proof: From Theorem 4.1 we know that for all optimal lines Med1 is true. It remains to show that

$$\sum_{Ex_m \in \mathcal{B}^-_{l^*}} w_m < \frac{1}{2}W \text{ and}$$

$$\sum_{Ex_m \in \mathcal{B}^+_{l^*}} w_m < \frac{1}{2}W$$

hold for all optimal lines l^*. Therefore we assume that this is not true for a median line $l^* = l_{s_1,s_2,b}$. Since l^* is pseudo-halving (see Theorem 3.3) we know that

$$\sum_{Ex_m \in \mathcal{B}^-_{l^*}} w_m \leq \frac{1}{2}W \text{ and}$$

$$\sum_{Ex_m \in \mathcal{B}^+_{l^*}} w_m \leq \frac{1}{2}W.$$

Hence we can assume without loss of generality that

$$\sum_{Ex_m \in \mathcal{B}_{l^*}^-} w_m = \frac{1}{2}W \text{ and } \sum_{Ex_m \in \mathcal{B}_{l^*}^+} w_m < \frac{1}{2}W.$$

Let $\mathcal{B}_{l^*}^- = \{(x_1, x_2) : s_1x_1 + s_2x_2 + b < 0\}$ and $\mathcal{M}^- = \{m \in \mathcal{M} : Ex_m \in \mathcal{B}_{l^*}^-\}$ the corresponding set of indices. Decreasing b to $b - \delta$ with sufficiently small $\delta > 0$ yields a line $l' = l_{s_1,s_2,b-\delta}$ which does not contain any of the existing facilities. For the objective value of l' we calculate

$$\begin{aligned} f(l') - f(l^*) &= \sum_{m \in \mathcal{M}^-} w_m C(-\delta) + \sum_{m \in \mathcal{M}^-} w_m C\delta \\ &= C\delta \left(- \sum_{m \in \mathcal{M}^-} w_m + \sum_{m \in \mathcal{M}^-} w_m + \sum_{m: Ex_m \in l^*} w_m \right) \\ &= C\delta \left(\frac{W}{2} - \frac{W}{2} \right) = 0. \end{aligned}$$

Here $C = C(-\frac{s_1}{s_2}, \gamma, d_{ver})$ is the constant defined in Lemma 3.10 that transfers the distance derived from γ to the vertical distance d_{ver} (if l^* is not a vertical line; for a vertical line l^* we can use $C = C(-\frac{s_1}{s_2}, \gamma, d_{hor})$ analogously.) Therefore the line l' is also optimal. But since l' is not passing through two of the existing facilities this leads to a contradiction to Theorem 4.1.

QED

4.2 The Strong Blockedness Property

Similar to the strong incidence property for median lines we now want to improve the weak blockedness property for center problems. Although the result is analogous to the result for the median problem of the previous section, the proof will use purely geometrical arguments. Therefore we prove Lemma 4.3 about line transversals and Corollary 4.3 that will be needed in Chapter 5. (Recall that an introduction to transversal theory has been given in Section 1.4, see page 17.)

Lemma 4.3 *Let $\mathcal{A}$ be a finite family of compact convex and smooth sets in $\mathbb{R}^2$ and let l be a line transversal for $\mathcal{A}$. If l does not touch at least three of the sets in $\mathcal{A}$ then there exists a line transversal l' with*

$$l' \cap int(A) \neq \emptyset \text{ for all } A \in \mathcal{A}.$$

Proof:

- If l does not touch any of the sets in $\mathcal{A}$ then $l' = l$ meets the interior of all sets in $\mathcal{A}$.

- Now suppose $l = l_{s_1,s_2,b}$ touches exactly one set $\tilde{A}$ which lies, without loss of generality, above l (or on the right hand side of l in the case that l is a vertical line). Then for a small $\epsilon > 0$ $l' = l_{s_1,s_2,b+\epsilon}$ is still a line transversal, but with $l' \cap int(A) \neq \emptyset$ for all $m \in \mathcal{M}$.

- Now let exactly two sets $\tilde{A}, \bar{A}$ touch the line l. If both sets lie on the same side of l we proceed as in the previous case. If $\tilde{A}$ and $\bar{A}$ are on different sides of l then choose

$$\begin{aligned} \tilde{x} &\in \tilde{A} \cap l \text{ and} \\ \bar{x} &\in \bar{A} \cap l \end{aligned}$$

and define l' as a small rotation of l about

$$\frac{\tilde{x} + \bar{x}}{2}$$

towards both sets $\tilde{A}$ and $\bar{A}$. For $\tilde{x} \neq \bar{x}$ the new line l' separates $\tilde{x}$ from some other points in $\tilde{A}$ and also $\bar{x}$ from some other points in $\bar{A}$ such that $l' \cap int(A) \neq \emptyset$ for all $A \in \mathcal{A}$. If, on the other hand, $\tilde{x} = \bar{x}$ we need the additional assumption that $\tilde{A}$ and $\bar{A}$ are smooth sets. Thus the new line l' cannot be a supporting line of these sets at $\tilde{x} = \bar{x}$. Hence we know that l' passes through the interior of both sets $\tilde{A}$ and $\bar{A}$ and also through the interior of all other sets $A \in \mathcal{A}$.

QED

Corollary 4.3 *Let $\mathcal{A}$ be a family of compact convex sets in $\mathbb{R}^2$ and let l be a line transversal for $\mathcal{A}$. If l does not touch three of the sets in $\mathcal{A}$ then either*

- *there exists a line transversal l' with*

$$l' \cap int(A) \neq \emptyset \text{ for all } A \in \mathcal{A} \text{ or}$$

- *there exists a line transversal l' which touches three of the sets in $\mathcal{A}$.*

Proof: If l touches none or exactly one of the sets in $\mathcal{A}$ or if l touches exactly two of the sets in $\mathcal{A}$ in at least two different points then there exists a line transversal l' which meets the interior of all sets in $\mathcal{A}$. This follows directly from the proof of Lemma 4.3, since for these cases the assumption that the sets in $\mathcal{A}$ are smooth has not been used. Now let l be a line transversal that touches exactly two of the sets in $\mathcal{A}$, e.g., $\tilde{A}$ and $\bar{A}$, in one common point $x \in l \cap \tilde{A} \cap \bar{A}$. Then we rotate l about this point x. We stop the rotation as soon as we get a line l' that either touches another set in $\mathcal{A}$ (this must happen since all sets in $\mathcal{A}$ are bounded) or if we get a line l' that meets the interior of one of the sets $\tilde{A}, \bar{A}$.

- In the first case l' touches one of the remaining sets in $\mathcal{A}$ and is still tangent to the sets $\tilde{A}$ and $\bar{A}$, such that l' touches three of the sets in $\mathcal{A}$.
- In the second case l' touches at most one of the sets in $\mathcal{A}$, such that we can apply the proof of the previous lemma again. This yields a line l' which passes through the interior of all sets in $\mathcal{A}$.

QED

Using Corollary 4.3 we will give an alternative proof of Cen1 for all distances derived from norms in the next Chapter. Now we can apply Lemma 4.3 to establish the strong blockedness property, namely that *all* center lines are at maximum weighted distance from three of the existing facilities if the distance d is derived from a smooth norm.

Theorem 4.2 *γ is a smooth norm if and only if for all instances of $1l/\mathbb{R}^2/\cdot/\gamma/\max$ the following holds: All center lines are at maximum (weighted) distance from three of the existing facilities.*

Proof:

$\Longrightarrow$: Let d be derived from a smooth norm γ and assume that l^* is an optimal line for $1l/\mathbb{R}^2/\cdot/\gamma/\max$ with objective value g^* but is not at maximum distance from three of the existing facilities. Using Lemma 1.6 we then know that l^* is a line transversal for $\mathcal{A}(g^*)$ with

$$\begin{aligned}
\mathcal{A}(g^*) &= \{A_m(g^*) : m \in \mathcal{M}\} \text{ and} \\
A_m(g^*) &= \{x : w_m d(Ex_m, x) \le g^*\} \\
&= \frac{g^*}{w_m} B + Ex_m,
\end{aligned}$$

where $B = \{x : \gamma(x) \le 1\}$ is the unit ball of the norm γ. Since B is a convex and smooth set, the sets $A_m(g^*)$ are also convex and smooth for all $m \in \mathcal{M}$. So we can apply Lemma 4.3. We note that l^* is at maximum distance from one of the existing facilities Ex_k, $k \in \mathcal{M}$, if and only if $d(Ex_k, l^*) = g^*$ which equivalently means that l^* touches $A_k(g^*)$. Our assumption therefore yields that l^* touches at most two of the sets in $\mathcal{A}(g^*)$. Lemma 4.3 then yields that there exists a line l' passing through the interior of all sets $A_m(g^*) \in \mathcal{A}(g^*)$. Consequently there exists $\lambda < 1$ such that l' also meets all the shrunk sets $A_m(\lambda g^*)$ which, using Lemma 1.6 again, leads to

$$g(l') \le \lambda g^* < g^*.$$

Thus l' has a better objective value than l^* which is a contradiction, since we assumed optimality for l^*.

$\Longleftarrow$: Now let γ be a non-smooth norm with unit ball B. Then there exists a $t \in \mathbb{R}^2$ with $\gamma(t) = 1$ such that there are at least two supporting lines of

B at t. As in the proof of Theorem 4.1 let $\mathcal{J} = [\underline{s}, \overline{s}]$ denote the interval of all slopes of supporting lines to B at t, and without loss of generality let $0 \in int(\mathcal{J})$.

Define the following set of $M = 3$ existing facilities, shown in Figure 4.2: $Ex_1 = (0,0)$, $Ex_2 = 2t$, and $Ex_3 = (c,0)$, where c is a sufficiently large real number. Furthermore choose $w_m = 1$ for $m = 1,2,3$. The three lines which are at maximum distance from all three existing facilities are the lines passing through two of the three points P_{12}, P_{23}, and P_{13} according to Theorem 3.4 on page 69. Thus we have to evaluate

- the line l_1 through t and $\frac{1}{2}Ex_3$,
- the line l_2 through t and $\frac{1}{2}Ex_3 + t$, and
- the linc l_3 through $\frac{1}{2}Ex_3$ and $\frac{1}{2}Ex_3 + t$.

Since $0 \in int(\mathcal{J})$ the lines l_1 and l_2 are supporting lines of B at t if c is sufficiently large, i.e.,

$$g(l_1) = g(l_2) = \gamma(t) = 1, \text{ and}$$

$$g(l_3) > 1 \text{ again, if } c \text{ is large enough.}$$

Finally the line l through t and Ex_3 satisfies also $g(l) = 1$, but since $d(Ex_3, l) = 0$, l_3 is at maximum distance from only two of the existing facilities.

QED

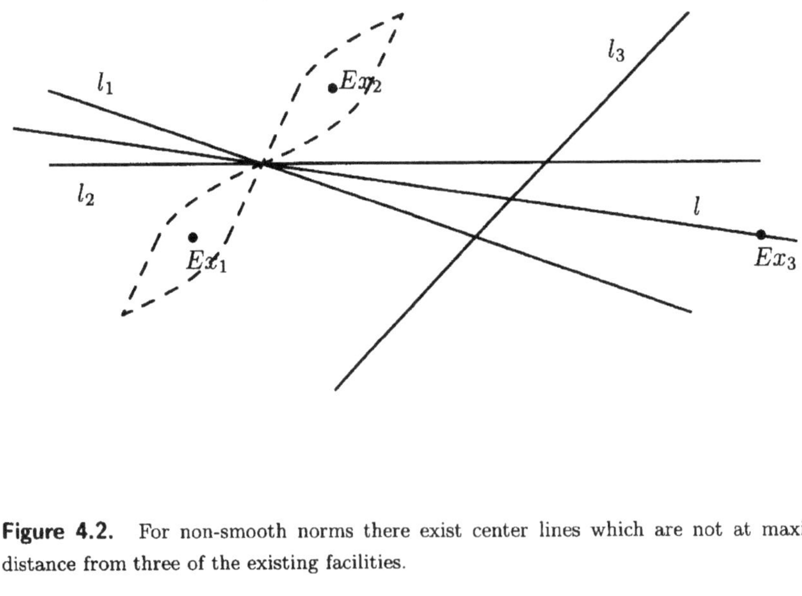

Figure 4.2. For non-smooth norms there exist center lines which are not at maximum distance from three of the existing facilities.

4.3 Determining all Optimal Lines

In this section we will describe how to find *all* optimal solutions of the line location problems $1l/\mathbb{R}^2/\cdot/\gamma/\sum$ and $1l/\mathbb{R}^2/\cdot/\gamma/\max$. The last two sections showed that this is easy if the distance measure is derived from a smooth norm γ: A simple enumeration of all candidate lines is possible using Algorithms 1, 2, or 3 of Section 3.2. On the other hand, for non-smooth norms γ, these algorithms will in general not find the whole set $\mathcal{L}^*$ of optimal lines, but only some optimal solutions. In the following we use the results of the last two sections to determine the set of all optimal lines $\mathcal{L}^*$ also for non-smooth norms. First, we need an additional definition.

Definition 20 *Let γ be a norm with unit ball B and let $\gamma(t) = 1$. t is called a* **corner point** *of the norm γ if there exist at least two supporting lines of B at t.*
The set of all corner points of γ will be denoted by $\mathcal{CP}_\gamma$.

Note that the norm γ is smooth if and only if $\mathcal{CP}_\gamma = \emptyset$, and that for block norms γ_B we have that $\mathcal{CP}_{\gamma_B} = Ext(B)$ is the set of extreme points of the unit ball B of the norm. Furthermore, since we discuss norms, the set of corner points is symmetric, i.e., $t \in \mathcal{CP}_\gamma \iff -t \in \mathcal{CP}_\gamma$. In the following we need only one of the two points t and $-t$, and therefore we assume that all symmetric counterparts have been dropped from $\mathcal{CP}_\gamma$. Using this notation we can state the main theorem of this section.

Theorem 4.3 *Let γ be an arbitrary norm with corner points $\mathcal{CP}_\gamma$.*

1. *If l^* is an optimal line for $1l/\mathbb{R}^2/\cdot/\gamma/\sum$, then l^* passes through two of the existing facilities, or there exists a $t \in \mathcal{CP}_\gamma$ such that $d_t(x, l^*) = d(x, l^*)$ for all $x \in \mathbb{R}^2$.*

2. *If l^* is an optimal line for $1l/\mathbb{R}^2/\cdot/\gamma/\max$, then l^* is at maximum distance from three of the existing facilities, or there exists a $t \in \mathcal{CP}_\gamma$ such that $d_t(x, l^*) = d(x, l^*)$ for all $x \in \mathbb{R}^2$.*

Proof: Let the distance d be derived from the norm γ.

1. Take an optimal line l^* such that $d_t(x, l^*) \neq d(x, l^*)$ for all $t \in \mathcal{CP}_\gamma$. Hence, if we replace d by a distance d_t (according to Lemma 3.12) we know that t is not a corner point of γ, i.e., there exists only one supporting line of B at t.

   Now assume that l^* does not pass through two of the existing facilities. Then we consider the problem with t-distance and find a line l' passing through two of the existing facilities and satisfying

$$\sum_{m\in\mathcal{M}} w_m d_t(Ex_m, l') \leq \sum_{m\in\mathcal{M}} w_m d_t(Ex_m, l^*) = f(l^*).$$

Using the same arguments as in the proof of Theorem 4.1 (i.e., we assume without loss of generality that l^* and l' have different slopes, see Lemma 4.2, and then derive a contradiction to $d(x,l') = d_t(x,l')$) we conclude that

$$f(l') < \sum_{m\in\mathcal{M}} w_m d_t(Ex_m, l'),$$

contradicting the optimality of l^*.

2. For the center objective we can proceed analogously, but using again the idea of the proof of Theorem 4.1 and not that of the geometric proof of Section 4.2. Note that the result of Lemma 4.2 for the center function will be proven later for the higher dimensional case, see Lemma 7.6 in Chapter 7.

QED

Theorem 4.3 leads to the following idea for the determination of the whole set $\mathcal{L}^*$. To find all median lines, one has to check all lines passing through two of the existing facilities (or which are at maximum distance from at least three of the existing facilities for solving center problems) and to solve all corresponding problems with distance d_t for all corner points t of γ. This will be made precise in the following.

Lemma 4.4 *Let $\mathcal{L}^*$ be the set of optimal lines for a line location problem (median or center) with distance d derived from a norm γ. Let $l^* \in \mathcal{L}^*$ and $t \in \mathbb{R}^2$ be given such that $d_t(x,l^*) = d(x,l^*)$ for all $x \in \mathbb{R}^2$. For the set $\mathcal{L}_t^*$ of optimal lines for the corresponding line location problem with t-distance then the following criteria hold.*

1. $l^* \in \mathcal{L}_t^*$.
2. $\mathcal{L}_t^* \subseteq \mathcal{L}^*$.
3. *All lines $l \in \mathcal{L}_t^*$ satisfy $d_t(x,l) = d(x,l)$ for all $x \in \mathbb{R}^2$.*

Proof: Take any line $l \in \mathcal{L}_t^*$. We evaluate the objective value of l. For the median function we obtain:

$$\begin{aligned} f(l) &= \sum_{m\in\mathcal{M}} w_m d(Ex_m, l) \\ &\le \sum_{m\in\mathcal{M}} w_m d_t(Ex_m, l), \text{ see Lemma 3.11} \\ &\le \sum_{m\in\mathcal{M}} w_m d_t(Ex_m, l^*) \text{ since } l \in \mathcal{L}_t^* \\ &= \sum_{m\in\mathcal{M}} w_m d(Ex_m, l^*) \text{ since } d_t(x,l^*) = d(x,l^*) \\ &= f(l^*), \end{aligned}$$

Since l^* is optimal, we conclude that $f(l) = f(l^*)$. Hence we get:

1. $\sum_{m \in \mathcal{M}} w_m d_t(Ex_m, l) = \sum_{m \in \mathcal{M}} w_m d_t(Ex_m, l^*)$ such that $l^* \in \mathcal{L}_t^*$.
2. $l \in \mathcal{L}^*$, yielding that $\mathcal{L}_t^* \in \mathcal{L}^*$.
3. $\sum_{m \in \mathcal{M}} w_m d(Ex_m, l) = \sum_{m \in \mathcal{M}} w_m d_t(Ex_m, l)$. Furthermore, according to Lemma 3.10 (see page 62) there exists a constant C such that $d_t(x, l) = Cd(x, l)$ for all $x \in \mathbb{R}^2$. (Note that $d_t(x, l) < \infty$ and therefore the slope of t is not equal to the slope of l.) Together, we conclude that $C = 1$ and hence that $d(x, l) = d_t(x, l)$ for all $x \in \mathbb{R}^2$ holds for all lines $l \in \mathcal{L}_t^*$.

For the center objective $g(l) = g(l^*)$ can be derived analogously to prove the second part of the lemma.

QED

To solve line location problems with t-distance we use Lemma 3.9 and further transform the problem to a problem with vertical distance. As these problems are convex, it is possible to determine the set of all optimal solutions in this case. We now discuss the structure of $\mathcal{L}^*$ for line location problems with vertical distance. To this end we need the following definition.

Definition 21 *Let $\mathcal{L}$ be a set of non-vertical lines and let $T(\mathcal{L}) = \{T(l) : l \in \mathcal{L}\}$ be their representation by points (s, b) in the dual space. Then the* **convex hull of the lines** *$l \in \mathcal{L}$ is given by*

$$conv(\mathcal{L}) := T^{-1}(conv(T(\mathcal{L}))).$$

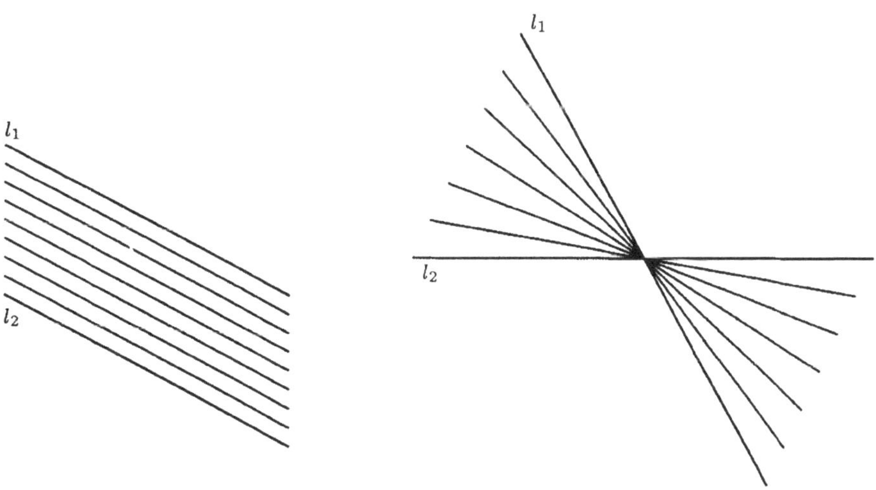

Figure 4.3. Determining the convex hull of two lines l_1 and l_2.

To determine the convex hull of two non-vertical lines l_1, l_2 we have to distinguish two cases shown in Figure 4.3.

The two lines are parallel : If l_1 and l_2 are parallel, i.e., $l_1 = l_{s,-1,b_1}$ and $l_2 = l_{s,-1,b_2}$ with $b_1 < b_2$, then we determine

$$conv(\{(s,b_1),(s,b_2)\}) = \{(s,b) : b_1 \leq b \leq b_2\}$$

and conclude that the convex hull of l_1 and l_2 is the set of all lines parallel to l_1 that lie between l_1 and l_2, i.e.,

$$conv(\{l_1,l_2\}) = \{l_{s,-1,b} : b_1 \leq b \leq b_2\}.$$

The two lines intersect : Let $\{x\} = l_1 \cap l_2$ with $x = (x_1, x_2)$. Then we can rewrite l_1 and l_2 as

$$\begin{aligned} l_1 = l_{s_1,-1,b_1} &= l_{s_1,-1,x_2-s_1x_1}, \\ l_2 = l_{s_2,-1,b_2} &= l_{s_2,-1,x_2-s_2x_1}. \end{aligned}$$

The convex hull of the corresponding points in the dual space is then given by

$$\begin{aligned} conv\ & (\{(s_1, x_2 - s_1x_1), (s_2, x_2 - s_2x_1)\}) \\ &= \{(s_2 + \lambda(s_1 - s_2), x_2 - (s_2 + \lambda(s_1 - s_2))x_1) : 0 \leq \lambda \leq 1\}. \end{aligned}$$

Assuming $s_1 \leq s_2$, $conv(\{l_1,l_2\})$ thus is the set of all lines $l_{s,-1,x_2-sx_1}$ with $s_1 \leq s \leq s_2$, i.e., all lines l in $conv(\{l_1,l_2\})$ pass through the same point x and lie between l_1 and l_2.

In the case that $\mathcal{L}$ consists of more than two lines it is not so easy to show $conv(\mathcal{L})$ in a picture. This is demonstrated in Figure 4.4. Here three lines l_1, l_2, and l_3 are given. The dashed lines graphed in that figure do *not* belong to $conv(\{l_1,l_2,l_3\})$!

Define

$$\begin{aligned} \mathcal{L}^*_{inc} &= \{l \in \mathcal{L}^* : l \text{ passes through two of the existing facilities}\} \text{ and} \\ \mathcal{L}^*_{blo} &= \{l \in \mathcal{L}^* : l \text{ is at maximum distance from three of the existing facilities}\} \end{aligned}$$

as the set of all median lines incident with two existing facilities and as the set of all blocked center lines. The corresponding distance function will be given individually in each case. With this notation we can formulate the following lemma to describe the set of optimal lines for line location problems with vertical distance.

Lemma 4.5 *Consider line location problems with vertical distance d_{ver}.*

1. *For the median problem we have $\mathcal{L}^* = conv(\mathcal{L}^*_{inc})$.*

2. *For the center problem we have $\mathcal{L}^* = conv(\mathcal{L}^*_{blo})$.*

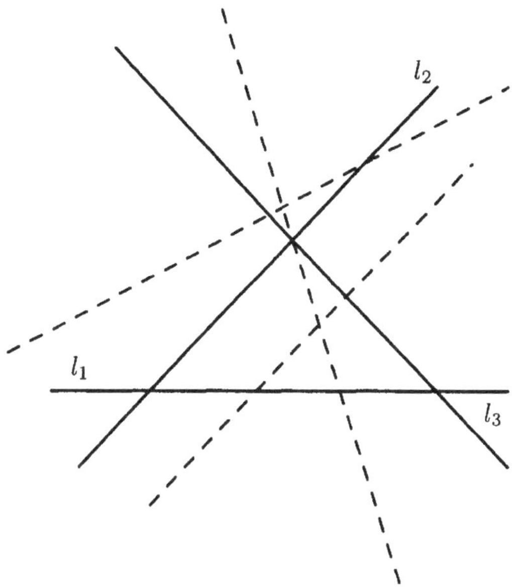

Figure 4.4. The dashed lines do *not* belong to the convex hull of l_1, l_2, and l_3.

Proof: We use the dual interpretation of line location problems. In the dual space $T(\mathcal{L}^*_{inc})$ and $T(\mathcal{L}^*_{blo})$ are the sets of optimal cell vertices for the median and the center problem, respectively. Since both problems are convex, Theorem 1.3 yields that the set of optimizers in the dual space is given by $conv(T(\mathcal{L}^*_{inc}))$ for the median problem and by $conv(T(\mathcal{L}^*_{blo}))$ for the center problem, respectively. Transforming these sets back to the primal space one sees that $T^{-1}(conv(T(\mathcal{L}^*_{inc})))$ and $T^{-1}(conv(T(\mathcal{L}^*_{blo})))$ are the sets of optimal lines for the original problems.

QED

Example 4.2 *Consider Example 4.1 (see also Figure 1.2 on page 16), but with vertical distance d_{ver}. Then $\mathcal{L}^*_{inc}$ is given by the four non-vertical lines passing through two of the existing facilities, i.e.,*

$$\mathcal{L}^*_{inc} = \{l_{12}, l_{13}, l_{24}, l_{34}\}.$$

The convex hull of $\mathcal{L}^*_{inc}$ is given by all lines l that meet both vertical facets of the square Q, i.e., both line segments between Ex_1 and Ex_4 and between Ex_2 and Ex_3. According to Lemma 4.5 this is the set of all median lines, i.e., $conv(\mathcal{L}^*_{inc}) = \mathcal{L}^*$.

The following corollary generalizes the results of Lemma 4.5 to the case of line location problems with t-distances.

Corollary 4.4 *Let* $t \in \mathbb{R}^2$.

1. *For the median problem with distance* d_t *we have* $\mathcal{L}^* = conv(\mathcal{L}^*_{inc})$.

2. *For the center problem with distance* d_t *we have* $\mathcal{L}^* = conv(\mathcal{L}^*_{blo})$.

Proof: Using Lemma 4.5 and a transformation D according to Lemma 3.9, the result is straightforward.

QED

Theorem 4.4 *Let d be a distance derived from the norm γ. Let $\mathcal{L}^*_{inc}$ and $\mathcal{L}^*_{blo}$ refer to the optimal solutions of the line location problem (median problem and center problem) with distance d. Using the notation*

$$\mathcal{L}(t) := \{ \text{ lines } l : d(x,l) = d_t(x,l) \text{ for all } x \in \mathbb{R}^2\},$$

the set of optimal lines $\mathcal{L}^*$ is given by the following formulas.

For the median problem:

$$\mathcal{L}^* = \left(\bigcup_{t \in CP_\gamma} conv(\mathcal{L}^*_{inc} \cap \mathcal{L}(t)) \right) \cup \mathcal{L}^*_{inc}.$$

For the center problem:

$$\mathcal{L}^* = \left(\bigcup_{t \in CP_\gamma} conv(\mathcal{L}^*_{blo} \cap \mathcal{L}(t)) \right) \cup \mathcal{L}^*_{blo}.$$

Proof: We prove Theorem 4.4 only for the median problem, since the proof for center problems can be derived analogously.

$\subseteq$: Let $l^* \in \mathcal{L}^*$. Theorem 4.3 yields that $l^* \in \mathcal{L}^*_{inc}$, or there exists a $t \in CP_\gamma$ with $d_t(x, l^*) = d(x, l^*)$ for all $x \in \mathbb{R}^2$. Lemma 4.4 then shows that

1. $l^* \in \mathcal{L}^*_t$,
2. $\mathcal{L}^*_t \subseteq \mathcal{L}^*$, and that
3. $\mathcal{L}^*_t \subseteq \mathcal{L}(t)$.

Now consider the subset of all optimal lines for the corresponding problem with distance d_t that pass through two of the existing facilities, i.e.,

$$\mathcal{L}^*_{inc}(d_t) = \{l \in \mathcal{L}^*_t : l \text{ passes through two of the existing facilities}\}.$$

Since $\mathcal{L}^*_{inc}(d_t) \subseteq \mathcal{L}^*_t$ we know that all lines $l \in \mathcal{L}^*_{inc}(d_t)$ also satisfy 2. and 3. and hence $\mathcal{L}^*_{inc}(d_t) \subseteq \mathcal{L}^*_{inc} \cap \mathcal{L}(t)$. Finally, Corollary 4.4 yields that

$$\mathcal{L}^*_t = conv(\mathcal{L}^*_{inc}(d_t)) \subseteq conv(\mathcal{L}^*_{inc} \cap \mathcal{L}(t)),$$

which shows that $l^* \in conv(\mathcal{L}^*_{inc} \cap \mathcal{L}(t))$.

$\supseteq$: For $l \in \mathcal{L}^*_{inc}$ obviously $l \in \mathcal{L}^*$. It is also clear that $(\mathcal{L}^*_{inc} \cap \mathcal{L}(t)) \subseteq \mathcal{L}^*$. Furthermore we know from the first part of Lemma 4.4 that $(\mathcal{L}^*_{inc} \cap \mathcal{L}(t)) \subseteq \mathcal{L}^*_t$. Since $\mathcal{L}^*_t = conv(\mathcal{L}^*_t)$, we have that also

$$conv(\mathcal{L}^*_{inc} \cap \mathcal{L}(t)) \subseteq \mathcal{L}^*_t.$$

Finally, if $\mathcal{L}^*_{inc} \neq \emptyset$, then the second part of Lemma 4.4 yields that $\mathcal{L}^*_t \subseteq \mathcal{L}^*$. Together we get that $conv(\mathcal{L}^*_{inc} \cap \mathcal{L}(t)) \subseteq \mathcal{L}^*$.

QED

Corollary 4.5 *Let d be a distance derived from the norm γ. Then the following statements hold.*

For the median problem :
If $|\mathcal{L}^*_{inc}| = 1$ or if $|\mathcal{L}^*_{inc} \cap \mathcal{L}(t)| = 1$ for all $t \in \mathcal{CP}_\gamma$ then $\mathcal{L}^* = \mathcal{L}^*_{inc}$.

For the center problem :
If $|\mathcal{L}^*_{blo}| = 1$ or if $|\mathcal{L}^*_{blo} \cap \mathcal{L}(t)| = 1$ for all $t \in \mathcal{CP}_\gamma$ then $\mathcal{L}^* = \mathcal{L}^*_{blo}$.

Finally we can formulate the following algorithms to determine all optimal solutions of line location problems with distance d derived from an arbitrary norm. The correctness of both approaches follows from the theory developed above.

Algorithm 5 (for solving $1l/\mathbb{R}^2/\cdot/\gamma/\sum$)

Input: Distance d, derived from a norm γ and $Ex_m, w_m > 0$ for all $m \in \mathcal{M}$

Output: All median lines $\mathcal{L}^*$ solving $1l/\mathbb{R}^2/\cdot/\gamma/\sum$

1. Determine the set $\mathcal{L}^*_{inc}$, e.g., by using Algorithm 1. Set $\mathcal{L}^* := \mathcal{L}^*_{inc}$.

2. If $|\mathcal{L}^*_{inc}| = 1$ or if γ is a smooth norm then Output: $\mathcal{L}^* = \mathcal{L}^*_{inc}$.

3. For all $t \in \mathcal{CP}_\gamma$

1. Determine

$$\mathcal{L}^t_{inc} := \{l \in \mathcal{L}^*_{inc} : d(x,l) = d_t(x,l) \text{ for all } x \in \mathbb{R}^2\}.$$

2. If $|\mathcal{L}^t_{inc}| > 1$ then $\mathcal{L}^* := \mathcal{L}^* \cup conv(\mathcal{L}^t_{inc})$.

4. Output: $\mathcal{L}^*$.

Note that for one line l there may exist t_1, t_2 with

$$d_{t_1}(x,l) = d(x,l) = d_{t_2}(x,l),$$

meaning that the sets $\mathcal{L}^t_{inc} = \mathcal{L}^*_{inc} \cap \mathcal{L}(t)$ are not necessarily disjoint. The same is true for the sets $\mathcal{L}^t_{blo} = \mathcal{L}^*_{blo} \cap \mathcal{L}(t)$ which will be used in the following algorithm for center problems.

Algorithm 6 (for solving $1l/\mathbb{R}^2/\cdot/\gamma/\max$)

Input: Distance d, derived from a norm γ and $Ex_m, w_m > 0$ for all $m \in \mathcal{M}$

Output: All center lines $\mathcal{L}^*$ solving $1l/\mathbb{R}^2/\cdot/\gamma/\max$

1. Determine the set $\mathcal{L}^*_{blo}$, e.g., by using Algorithm 2. Set $\mathcal{L}^* := \mathcal{L}^*_{blo}$.

2. If $|\mathcal{L}^*_{blo}| = 1$ or if γ is a smooth norm then Output: $\mathcal{L}^* = \mathcal{L}^*_{blo}$.

3. For all $t \in \mathcal{CP}_\gamma$

1. Determine

$$\mathcal{L}^t_{blo} := \{l \in \mathcal{L}^*_{blo} : d(x,l) = d_t(x,l) \text{ for all } x \in \mathbb{R}^2\}.$$

2. If $|\mathcal{L}^t_{blo}| > 1$ then $\mathcal{L}^* := \mathcal{L}^* \cup \mathcal{L}^t_{blo}$.

4. Output: $\mathcal{L}^*$.

5
LINE LOCATION WITH OTHER DISTANCES

5.1 Results for Gauges

If the unit ball B of the distance measure is a convex set containing the origin that is not symmetric, B does not define a norm, but a gauge (see Definition 2). For such unit balls we use the notation $\tilde{B}$, and we use $\tilde{\gamma}$ for the corresponding gauge function. The derived distance function d is defined as

$$d(x, y) = \tilde{\gamma}(y - x),$$

but note that in general this is not equal to $\tilde{\gamma}(x - y)$. Consequently, $d(x, y) = d(y, x)$ is in general not true if d is derived from a gauge. For this reason Lemma 3.10 of Section 3.2 (see page 62) is not true. Therefore, also Lemma 3.12 (see page 65) does not hold, i.e., in general we cannot replace the distance function d by a t-distance d_t. But we can relax Lemma 3.10 to the following result.

Lemma 5.1 *Let d be derived from a gauge $\tilde{\gamma}$. Let $s \in \mathbb{R} \cup \{\infty\}$ be a slope and let $t \in \mathbb{R}^2$ be such that t does not have the slope s. Then there exist two real constants $C^+ = C^+(s, d, d_t)$ and $C^- = C^-(s, d, d_t)$, $C^+, C^- > 0$ such that for any line l with slope s the following conditions hold.*

$$\begin{aligned} d(x, l) &= C^+ \; d_t(x, l) \text{ for all } x \in \mathcal{B}_l^+ \text{ and} \\ d(x, l) &= C^- \; d_t(x, l) \text{ for all } x \in \mathcal{B}_l^-. \end{aligned}$$

(Recall that $\mathcal{B}_l^+$ and $\mathcal{B}_l^-$ are the two open halfplanes separated by the line l without the line itself, as introduced in Definition 8 in Section 1.3.)

Proof: Rereading the proof of Lemma 3.10 (see page 62) we note that for gauges $\tilde{\gamma}$ equation 3.6, namely

$$\tilde{\gamma}(|\beta|y) = \tilde{\gamma}(\beta y)$$

is in general not true. But the proof still holds for $\beta \geq 0$. Therefore we start again with $x = 0$, but with two lines l_{s_1,s_2,b_0^+}, l_{s_1,s_2,b_0^-} with $b_0^+ > 0$ and $b_0^- < 0$. Thus we get two constants $C^+, C^- > 0$. Then for all lines $l = l_{s_1,s_2,b}$ with $s = -\frac{s_1}{s_2}$ and $b > 0$ we get $\frac{b}{b_0^+} \geq 0$, and therefore

$$d(0, l_{s_1,s_2,b}) = \frac{b}{b_0^+} d(0, l_{s_1,s_2,b_0^+}) = \frac{b}{b_0^+} C^+ l_2(0, l_{s_1,s_2,b_0^+}) = C^+ l_2(0, l_{s_1,s_2,b}).$$

On the other hand, we get that $\frac{b}{b_0^-} \geq 0$ for lines $l = l_{s_1,s_2,b}$ with $s = -\frac{s_1}{s_2}$ and $b < 0$. Analogously we can derive

$$d(0, l_{s_1,s_2,b}) = C^- l_2(0, l_{s_1,s_2,b}).$$

For $x = 0$ this yields that we have to use C^+ for lines lying above x and C^- for lines lying below x. Since gauges are translation invariant in the same way as norms, this is also true for all $x \in \mathbb{R}^2$.

QED

Lemma 5.2 *Let d be derived from a gauge $\tilde{\gamma}$. Then there exists a median line that passes through one of the existing facilities.*

Proof: Let l^* be an optimal line passing through none of the existing facilities. Choose t such that the slope of t is not equal to the slope of l^*. Without loss of generality assume that the slope s of l^* is not infinite, and therefore choose $t = e_2$. (If $s = \infty$ rotate the coordinate system.) Then, according to Lemma 5.1, there exist constants $C^+, C^- > 0$ such that

$$\begin{aligned} d(x, l^*) &= C^+ \, d_{ver}(x, l^*) \text{ for all } x \in \mathcal{B}_{l^*}^+ \text{ and} \\ d(x, l^*) &= C^- \, d_{ver}(x, l^*) \text{ for all } x \in \mathcal{B}_{l^*}^-. \end{aligned}$$

It follows that

$$\begin{aligned} f(l^*) &= \sum_{m \in \mathcal{M}} w_m d(Ex_m, l^*) \\ &= \sum_{Ex_m \in \mathcal{B}_{l^*}^+} w_m C^+ d_{ver}(Ex_m, l^*) + \sum_{Ex_m \in \mathcal{B}_{l^*}^-} w_m C^- d_{ver}(Ex_m, l^*) \\ &= \sum_{m \in \mathcal{M}} w'_m d_{ver}(Ex_m, l^*) \\ &= \sum_{m \in \mathcal{M}} w'_m |a_{m1} s - a_{m2} + b|, \end{aligned}$$

$$\text{where} \quad w'_m = \begin{cases} C^+ w_m & \text{if } Ex_m \in \mathcal{B}_{l^*}^+ \\ C^- w_m & \text{if } Ex_m \in \mathcal{B}_{l^*}^-. \end{cases}$$

The line l^* contains none of the existing facilities and thus $|a_{m1}s - a_{m2} + b| \neq 0$ for all $m \in \mathcal{M}$. Fixing the slope of l^* but translating the line leads to the one-dimensional piecewise linear convex function

$$\bar{f}(b) = \sum_{m \in \mathcal{M}} w'_m |a_{m1}s - a_{m2} + b|$$

with breakpoints $a_{m2} - a_{m1}s$. Lemma 1.9 implies that there exists a breakpoint $b^0 = a_{k2} - sa_{k1}$ for $k \in \mathcal{M}$, such that

$$\bar{f}(b^0) \leq \bar{f}(b)$$

and such that there is no other breakpoint between b and b^0. Therefore $l^0 = l_{s,-1,b^0}$ is passing through Ex_k and

$$f(l_{s_1,s_2,b^0}) = \bar{f}(b^0) \leq \bar{f}(b) = f(l^*).$$

QED

Lemma 5.2 cannot be strengthened to Med1 as the following counterexample shows.

Example 5.1 *Let the gauge $\tilde{\gamma}$ be given by its unit ball*

$$\tilde{B} = conv\,\{(1,1), (-1,1), (0,-2)\},$$

and let $Ex_1 = (1,1)$, $Ex_2 = (-1,1)$, and $Ex_3 = (0,-2)$. Furthermore, $w_m = 1$ for $m = 1,2,3$ (see Figure 5.1). Then we consider the lines l_{12} through Ex_1 and Ex_2, l_{13} through Ex_1 and Ex_3 and l_{23} through Ex_2 and Ex_3.

$$\begin{aligned} f(l_{12}) &= d(Ex_3, l_{12}) = 3, \\ f(l_{13}) &= d(Ex_2, l_{13}) = 3, \text{ and} \\ f(l_{23}) &= f(l_{13}) = 3, \end{aligned}$$

but with $l^* = \{(x_1, x_2) : x_1 = 0\}$ we can calculate that

$$\begin{aligned} f(l^*) &= d(Ex_1, l^*) + d(Ex_2, l^*) + d(Ex_3, l^*) \\ &= 1 + 1 + 0 \\ &= 2, \end{aligned}$$

such that we can conclude that there exists no optimal line passing through two of the existing facilities.

The next example shows that also Med2 does not hold in general if d is derived from a gauge.

Example 5.2 *Let another polyhedral gauge $\tilde{\gamma}$ be given by the unit ball*

$$\tilde{B} = conv\,\{(1,5), (-1,5), (0,-1)\}$$

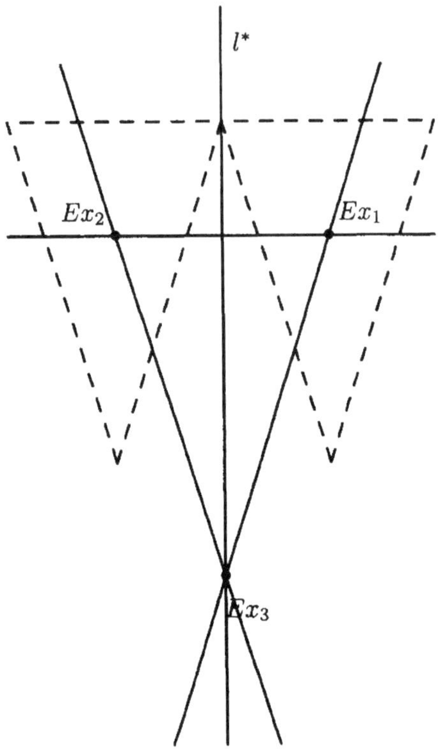

Figure 5.1. For gauges Med1 does not hold in general.

and $Ex_1 = (3,0)$*,* $Ex_2 = (-3,0)$*, and* $Ex_3 = (0,-5)$ *with weights* $w_1 = 1$*,* $w_2 = 1$*, and* $w_3 = 3$ *(see Figure 5.2). Then* $l^* = \{(x_1, x_2) : x_2 = 0\}$ *leads to*

$$f(l^*) = 3d(Ex_3, l^*) = 3,$$

and all lines l *not passing through*

$$\begin{aligned} A_1 &= \{x : \tilde{\gamma}(Ex_1, x) \leq 3\} = Ex_1 + 3\tilde{B} \text{ and} \\ A_2 &= \{x : \tilde{\gamma}(Ex_2, x) \leq 3\} = Ex_2 + 3\tilde{B} \end{aligned}$$

have objective value $f(l) > 3$*. But none of the lines passing through both* A_1 *and* A_2 *is pseudo-halving, yielding that no pseudo-halving line is optimal.*

Summarizing the last two examples we can conclude that, in general, Med1 and Med2 do not hold for distances derived from gauges $\tilde{\gamma}$. However, the properties

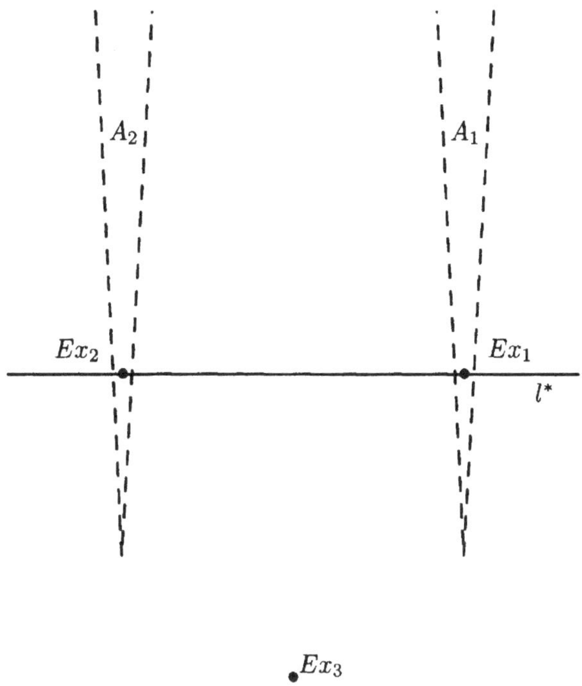

Figure 5.2. Med2 does in general not hold for gauges.

Cen1 and Cen2 are both true in the case of gauges $\tilde{\gamma}$. To show this we use Lemma 1.6 given on page 20 in Section 1.4.

Lemma 5.3 *Let d be derived from a gauge $\tilde{\gamma}$. Then the following criteria hold.*

Cen1: *There exists a center line which is at maximum distance from three of the existing facilities.*

Cen2: *If all weights are equal there exists a center line which is parallel to a facet of the convex hull of the existing facilities.*

Proof:

Cen1: Let d be derived from a gauge $\tilde{\gamma}$ and assume that l^* is an optimal line for $1l/\mathbb{R}^2/\cdot/\tilde{\gamma}/\max$ with objective value g^* but that it is not at maximum distance from three of the existing facilities. Using Lemma 1.6 we know that l^* is a line transversal for $\mathcal{A}(g^*)$ with

$$\mathcal{A}(g^*) \quad = \quad \{A_m(g^*) : m \in \mathcal{M}\} \text{ and}$$

$$\begin{aligned} A_m(g^*) &= \{x : w_m \tilde{\gamma}(Ex_m - x) \le g^*\} \\ &= \frac{g^*}{w_m} \tilde{B} + Ex_m, \end{aligned}$$

where $\tilde{B} = \{x : \tilde{\gamma}(x) \le 1\}$ is the unit ball of the gauge $\tilde{\gamma}$. As $\tilde{B}$ is convex, the sets $A_m(g^*)$ are also convex for all $m \in \mathcal{M}$. Furthermore, since l^* is a line transversal for the family $\mathcal{A}(g^*)$ which touches at most two of the sets in $\mathcal{A}(g^*)$ (compare the proof of Theorem 4.2 in the previous Chapter) we can in the following apply Corollary 4.3 and get the following two cases.

- Either there exists a line l' passing through the interior of all sets $A \in \mathcal{A}(g^*)$. This line has a strictly smaller objective value than l^* (compare again the proof of Theorem 4.2 in the last Chapter), contradicting the optimality of l^*.
- Or there exists a line transversal l' touching three of the sets $A \in \mathcal{A}(g^*)$. Since l' is a line transversal for the family $\mathcal{A}(g^*)$ we know that $g(l') \le g^*$ (according to Lemma 1.6) such that l' is optimal and at maximum distance from at least three of the existing facilities.

Cen2 : If the weights are all equal to 1, then the sets $A_m(g^*)$ are all translates of the unit ball $\tilde{B}$ of the gauge $\tilde{\gamma}$. Using Cen1 we know that there exists an optimal line l^* which is a line transversal for $\mathcal{A}(g^*)$ and touches three of the sets $A_k(g^*), A_l(g^*), A_r(g^*) \in \mathcal{A}(g^*)$. Assume without loss of generality that $A_k(g^*)$ and $A_l(g^*)$ lie in the same halfplane $\mathcal{B}_{l^*}^+$ separated by the line l^*. Since both sets are translates we know that there exists a line l parallel to l^* passing through Ex_k and Ex_l. As l^* meets all sets $A \in \mathcal{A}(g^*)$, we conclude for all $m \in \mathcal{M}$ that

$$Ex_m \in \mathcal{B}_l^- \cup l,$$

i.e., all existing facilities lie in the same halfplane separated by the line l or on the line itself, such that l contains a facet of $conv(\mathcal{E}x)$.

QED

5.2 Results for Metrics

Now we discuss distance measures d which are not necessarily derived from a norm, but which are metrics. For the examples in this section we will use metrics which are derived from the rectangular distance l_1. Before we introduce these metrics, we need the following definition.

Definition 22 *A sequence $z_0, z_1, \ldots, z_L$ of points in the plane, $z_i \in \mathbb{R}^2$ for $i = 0, \ldots, L$, is called a* **rectangular path** *from x to y if the following two conditions are satisfied.*

1. *$z_0 = x$ and $z_L = y$, and*

2. *Any two consecutive points z_l and z_{l+1}, $l = 0, 1, \ldots, L-1$ are connected by some horizontal or vertical line segment.*

Let $\mathcal{R}_{x,y}$ denote the set of all rectangular paths from x to y.

We now construct a metric in the following way. We choose a finite set $\mathcal{H}w$ of vertical (or horizontal) line segments, halflines, or lines $h_1, h_2, \ldots, h_K$ (called **highways**) Within these highways we allow a shorter distance measure, i.e., it may be possible to travel faster along a highway $h_k \in \mathcal{H}w$. Formally, this distance measure can be defined as follows.

Definition 23 *Let $\mathcal{H}w = \{h_1, h_2, \ldots, h_K\}$ be a finite set of vertical or horizontal line segments, lines, or halflines, and for $k = 1, 2, \ldots, K$ let*

$$df_k : h_k \times h_k \to \mathbb{R}$$

be K functions such that for $k = 1, 2, \ldots, K$ the following relations hold for all $x, y \in h_k$:

1. $df_k(x, y) \leq l_1(x, y)$,

2. $df_k(x, y) = df_k(y, x)$,

3. $df_k(x, y) \geq 0$, *and*

4. $df_k(x, y) = 0 \Longleftrightarrow x = y$.

Then for all $x, y \in \mathbb{R}^2$ define

$$d(x, y) = \min_{P \in \mathcal{R}_{x,y}} d(P),$$

where we define for $P = z_0 z_1 \ldots z_L \in \mathcal{R}_{x,y}$

$$d(P) = \sum_{l=0}^{L-1} d(z_l, z_{l+1})$$

and

$$d(z_l, z_{l+l}) = \begin{cases} df_k(x, y) & \text{if } z_l, z_{l+1} \in h_k \\ l_1(x, y) & \text{otherwise} \end{cases}$$

We call d the **Manhattan metric with highways** $h_1, h_2, \ldots, h_K$.

Approaches to construct inhomogeneous spaces with differentiable distance functions can be found in Chapter 4 of [Fli97].

Lemma 5.4 *Let d be a Manhattan metric with highways as described in Definition 23. Then d is a metric.*

Proof:

1. $d(x,y) \geq 0$ and $d(x,y) = 0 \iff x = y$ is obvious, because it holds for l_1 and for df_k, $k = 1, 2, \ldots, K$.

2. $d(x,y) = d(y,x)$ also holds, since $l_1(x,y) = l_1(y,x)$, $df_k(x,y) = df_k(y,x)$ for $k = 1, 2, \ldots, K$ and, additionally,

$$z_0 z_1 \ldots z_L \in \mathcal{R}_{x,y} \iff z_L z_{L-1} \ldots z_0 \in \mathcal{R}_{y,x}.$$

   Thus for each path $P \in \mathcal{R}_{x,y}$ there exists a path $P' \in \mathcal{R}_{y,x}$ such that $d(P) = d(P')$.

3. To show that the triangle inequality $d(x,z) \leq d(x,y) + d(y,z)$ holds for all $x, y, z \in \mathbb{R}^2$ we take a minimal path $P_1 \in \mathcal{R}_{x,y}$ from x to y and a minimal path $P_2 \in \mathcal{R}_{y,z}$ from y to z. We define the rectangular path P from x to z as the sequence of points of P_1 enlengthened by the sequence of points of P_2. Then we get

$$d(x,z) \leq d(P) = d(P_1) + d(P_2) = d(x,y) + d(y,z),$$

   which completes the proof.

QED

Now we discuss line location problems with arbitrary metrics d. First, the following example shows that Med1 and Med2 do not hold in general for the median problem $1l/\mathbb{R}^2/\cdot/d/\sum$. We can even have the situation that no optimal line contains any of the existing facilities!

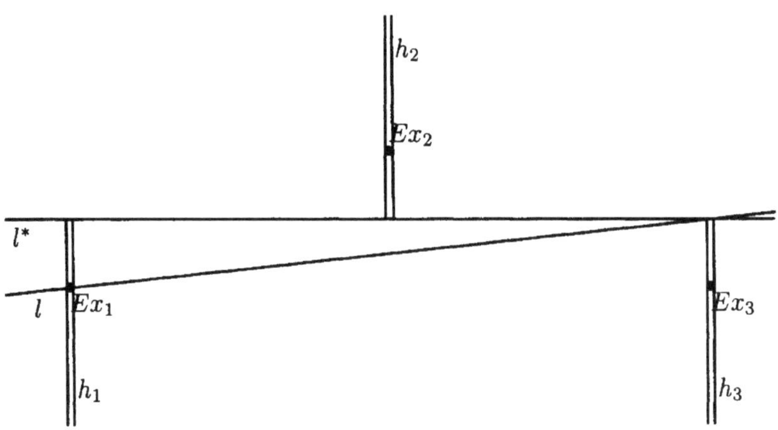

Figure 5.3. Med1 and Med2 do not hold for metrics in general. Here the median line l^* does not even contain any of the existing facilities.

Example 5.3 *For this example, see Figure 5.3. We consider the Manhattan metric with three highways,*

$$\begin{aligned} h_1 &= \{x = (x_1, x_2) : x_1 = -10 \text{ and } x_2 \leq 0\} \\ h_2 &= \{x = (x_1, x_2) : x_1 = 0 \text{ and } x_2 \geq 0\} \\ h_3 &= \{x = (x_1, x_2) : x_1 = 10 \text{ and } x_2 \leq 0, \} \end{aligned}$$

and the corresponding distance functions

$$df_k(x, y) = \frac{1}{3} l_1(x, y) \text{ for all } x, y \in h_k, k = 1, 2, 3.$$

The set of existing facilities $\mathcal{E}x$ is given by

$$Ex_1 = (-10, -2), \; Ex_2 = (0, 2), \; \text{and } Ex_3 = (10, -2).$$

Furthermore, define $w_m = 1, m = 1, 2, 3$. Then the two lines passing through the existing facilities Ex_1 and Ex_2, and through Ex_3 and Ex_2 both have objective value strictly greater than 8, and the line l' passing through Ex_1 and Ex_3 satisfies

$$f(l') = d(Ex_2, l') = \frac{1}{3}2 + 2 > 2.$$

But for the line $l^* = l_{0,-1,0}$ (the x-axis) we calculate

$$f(l^*) = \sum_{m=1}^{3} d(Ex_m, l^*) = \frac{1}{3}2 + \frac{1}{3}2 + \frac{1}{3}2 = 2.$$

Thus l^* has smaller objective value than all lines containing two of the existing facilities. It can be verified that all lines l containing at least one of the existing facilities have objective value

$$f(l) \geq 2 * \frac{2}{3} + 1,$$

compare line l in Figure 5.3. Hence, in this example no optimal line is passing through any of the existing facilities. Since l^* is not pseudo-halving, this is also a counterexample for Med2.

For the center problem with arbitrary metrics, Cen2 is also not true in general as we can see from the next example.

Example 5.4 *In this example we define one single highway*

$$h = \{x = (x_1, x_2) : x_1 = 0\} \text{ with } df(x, y) = \frac{1}{2} l_1(x, y)$$

as distance function for all $x, y \in h$. Furthermore, we choose

$$Ex_1 = (0, 0), Ex_2 = (0, 3), \; \text{and } Ex_3 = (8, 0),$$

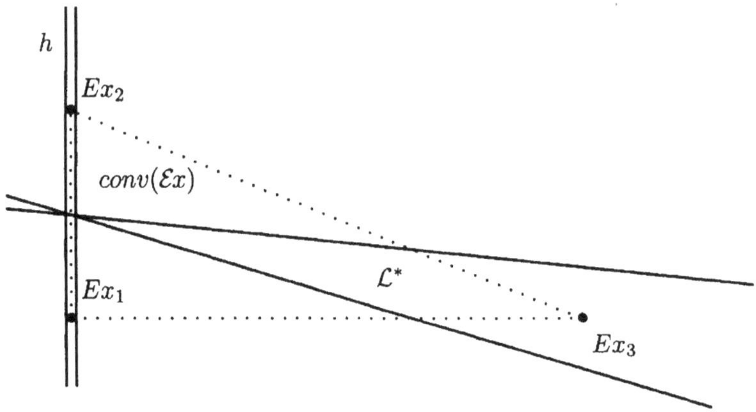

Figure 5.4. Cen2 does not hold for metrics in general.

and weights equal to 1 for all three existing facilities (see also Figure 5.4). Then for the center problem with this metric d, i.e., for $1l/\mathbb{R}^2/\cdot/d/\max$, *we get that exactly all lines l*

- *passing through* $(0, \frac{3}{2})$ *and*
- *satisfying* $d_{ver}(Ex_3, l) \leq \frac{3}{4}$

are optimal. None of them is parallel to a facet of the convex hull of the existing facilities.

Surprisingly we can confirm Cen1 for metrics with the following additional property.

Definition 24 *Let d be a metric in the plane, i.e.,* $d : \mathbb{R}^2 \times \mathbb{R}^2 \to \mathbb{R}$. *Then d is a* **strictly monotone metric** *if for all* $x \in \mathbb{R}^2$ *and for all directions* $t \in \mathbb{R}^2$ *the function* $f_{x,t} : \mathbb{R} \to \mathbb{R}$

$$f_{x,t}(\alpha) = d(x, x + \alpha t)$$

is strictly monotone on $\mathbb{R}^+ = \{\alpha \in \mathbb{R} : \alpha \geq 0\}$.

Note that all metrics derived from norms are strictly monotone and that also the two metrics defined in Examples 5.3 and 5.4 are strictly monotone metrics.

Lemma 5.5 *Let d be a strictly monotone metric. Then there exists an optimal line which is at maximum distance from three of the existing facilities.*

Proof: Let d be a metric and assume that l^* is an optimal line for $1l/\mathbb{R}^2/\cdot/d/\max$ with objective value g^* but that it is not at maximum distance from three of the existing facilities. Using the same notation as in the proof of Lemma 5.3 we define

$$A_m(g^*) \quad = \quad \{x : w_m d(Ex_m, x) \leq g^*\}$$

and use Lemma 1.6 to conclude that l^* is a line transversal for $\mathcal{A}(g^*)$ which touches at least two of the sets in $\mathcal{A}$. Applying Lemma 1.5 we conclude that l^* is also a line transversal for the family

$$\mathcal{A}_{conv} = \{conv(A) : A \in \mathcal{A}(g^*)\},$$

which touches at least two of the sets in $\mathcal{A}_{conv}$. Now Corollary 4.3 yields two cases.

Case 1: There exists a line l' passing through the interior of $conv(A_m(g^*))$ for all $m \in \mathcal{M}$. In this case let Ex_k be a facility which has maximum distance to l^*, and let $y \in l'$ be a point in the interior of $conv(A_k(g^*))$. Furthermore, define $t = Ex_k - y$. Then there exists an $\alpha \in \mathbb{R}$ with $|\alpha| > 1$ such that $\alpha t \in \delta(conv(A_k(g^*)))$. Hence we get

$$\begin{aligned} d(Ex_k, l') &\leq d(Ex_k, y) \\ &= d(Ex_k, Ex_k + t) \\ &= f_{Ex_k,t}(1) \\ &< f_{Ex_k,t}(|\alpha|) \\ &= d(Ex_k, l^*), \end{aligned}$$

because d is a strictly monotone metric, l^* touches $A_k(g^*)$ and αt lies on the boundary of $conv(A_k(g^*))$. Since this can be done for all existing facilities which are at maximum distance from l^* we conclude that $g(l') < g(l^*)$, contradicting the optimality of l^*.

Case 2: There exists a line transversal l' touching three of the sets $conv(\mathcal{A}(g^*))$. According to Lemma 1.5 (on page 19), l' touches three of the sets in $\mathcal{A}(g^*)$. Since l' is a line transversal for the family $\mathcal{A}(g^*)$ we know that $g(l') \leq g^*$ (using Lemma 1.6) so that l' is optimal and at maximum distance from at least three of the existing facilities.

QED

Note that for metrics that are not strictly monotone Cen1 does not hold any more. This is shown in the final example of this section.

Example 5.5 *In our final example with metrics (see Figure 5.5), we use again a Manhattan metric with one highway*

$$h = \{x = (x_1, x_2) : x_1 = 8\}.$$

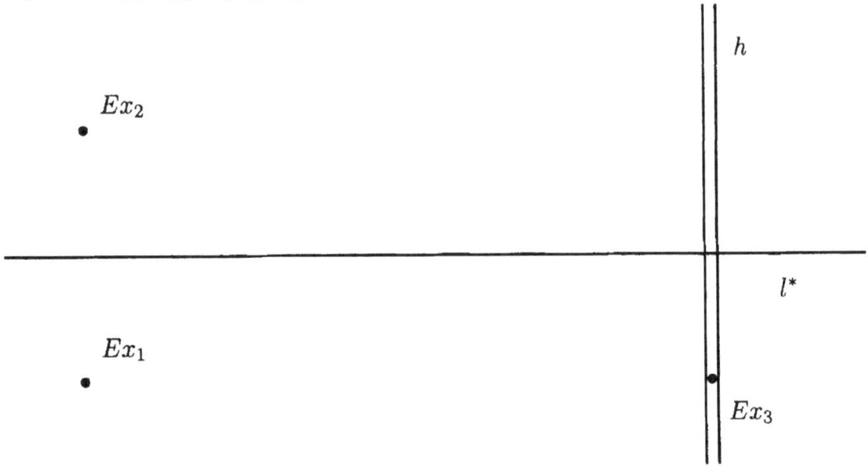

Figure 5.5. Cen1 does not hold for arbitrary metrics.

Furthermore we define

$$df(x, y) = \lfloor |x_2 - y_2| \rfloor \text{ for all } x, y \in h,$$

where $\lfloor x \rfloor = \min\{z : z \text{ is integer and } z \leq x\}$. *This metric is not strictly monotone, because for* $x = (8, 0)$ *and* $t = (0, 1)$ *we get that for* $\alpha \geq 0$

$$f_{x,t}(\alpha) = d(x, x + \alpha t) = d((8, 0), (8, \alpha)) = \lfloor \alpha \rfloor,$$

which is not a strictly monotone function.

With the same set of existing facilities as in Example 5.4 and with $w_m = 1, m = 1, 2, 3,$ *we observe that the best possible value for the center objective function is* $g^* = \frac{3}{2}$, *and an optimal line is for example*

$$l^* = \{x = (x_1, x_2) : x_2 = \frac{3}{2}\}.$$

But since $d(Ex_3, l)$ *is integer, no line* l' *with slope* ≤ 1 *satisfies*

$$d(Ex_m, l') = \frac{3}{2} \text{ for all } m = 1, 2, 3.$$

5.3 Results for Mixed Distance Functions

In this section we discuss a slightly modified version of the line location problem, namely we allow that each existing facility has its own distance measure. First we require that all these distance measures are derived from norms, i.e., to each existing facility a norm

$$\gamma_1, \gamma_2, \ldots, \gamma_M \text{ with derived distance } d_1, d_2, \ldots, d_M$$

is given. Now we want to find a line l such that we minimize either

the sum objective $f(l) = \sum_{m \in \mathcal{M}} w_m d_m(Ex_m, l)$, or

the center objective $g(l) = \max_{m \in \mathcal{M}} w_m d_m(Ex_m, l)$.

We classify problems of this type as $1l/\mathbb{R}^2/\cdot/\gamma_m/\sum$ and $1l/\mathbb{R}^2/\cdot/\gamma_m/\max$, respectively. Note that, if γ is a norm, also $\alpha\gamma$ is a norm for all $\alpha \in \mathbb{R}, \alpha > 0$. Hence weights w_m can be interpreted as factors changing the norms from γ_m to $w_m\gamma_m$. Therefore the additional consideration of weights is redundant in this context. Furthermore, this observation shows that Med2 and Cen2 cannot hold for $1l/\mathbb{R}^2/\cdot/\gamma_m/\sum$ and $1l/\mathbb{R}^2/w_m = 1/\gamma_m/\max$ in general:

- Suppose l is pseudo-halving for an instance of $1l/\mathbb{R}^2/\cdot/\gamma_m/\sum$ (for which not all existing facilities lie on the same line). Then take an existing facility $Ex_k \notin l$ and define

$$w'_k := Qw_k \text{ and } \gamma'_k := \frac{1}{Q}\gamma_k \text{ for } Q \in \mathbb{R}, Q > 0.$$

  This does not affect the set of existing facilities and the set of optimal lines. If we choose Q sufficiently large, l is not pseudo-halving for the new instance of $1l/\mathbb{R}^2/\cdot/\gamma_m/\sum$.

- Take a set of weighted existing facilities with one identical norm γ for which Cen2 is not true. Then define

$$\gamma'_m := w_m\gamma \text{ and } w'_m = 1 \text{ for all } m \in \mathcal{M}.$$

  Again, this does neither change the set $\mathcal{E}x$ nor the set of optimal lines, such that this constructs an instance of $1l/\mathbb{R}^2/w_m = 1/\gamma_m/\max$ for which Cen2 is not satisfied.

Also Med1 is not true in general for line location problems with mixed norms, as the following example shows.

Example 5.6 *We use the same three existing facilities as in Example 5.1, but modify the distance functions in the following way, see Figure 5.6.*

$$\begin{aligned}
\gamma_1 &= ext\{(-1,1),(0,-2),(1,-1),(0,2)\},\\
\gamma_2 &= ext\{(1,1),(0,-2),(-1,-1),(0,2)\}, \text{ and}\\
\gamma_3 &= ext\{(1,1),(1,-1),(-1,-1),(-1,1)\}.
\end{aligned}$$

Then none of the values $f(l_{12}), f(l_{13}), f(l_{23})$, and $f(l^*)$ changes compared with Example 5.1, because the new unit balls still touch the above lines in the same corner points. Consequently this is an example in which no median line contains two of the existing facilities.

However, the optimal line l^* in Example 5.6 contains one of the existing facilities. This is true in general for line location problems with mixed norms.

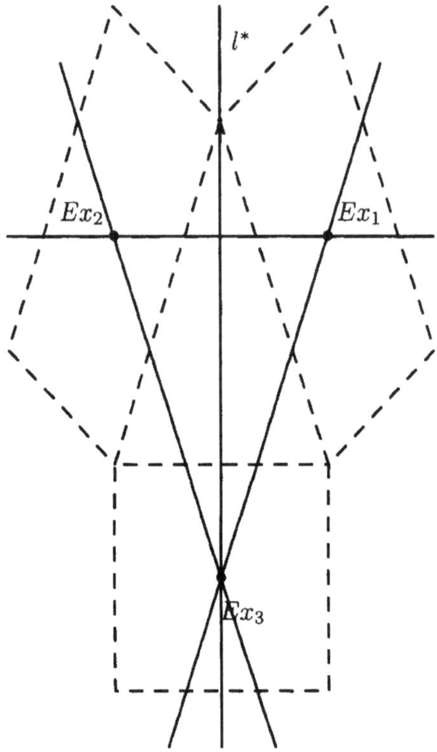

Figure 5.6. Med1 does not hold for mixed norms in general.

Lemma 5.6 *Let $\gamma_1, \gamma_2, \ldots, \gamma_M$ be norms. Then for the line location problem $1l/\mathbb{R}^2/\cdot/\gamma_m/\sum$ there exists a median line passing through one of the existing facilities.*

<u>Proof:</u> Let l^* be an optimal line passing through none of the existing facilities. Without loss of generality we assume that $l^*_{s^*,-1,b^*}$ is not vertical and set $t = e_2$. For all $\gamma_m, m \in \mathcal{M}$, we can derive constants

$$C_m = C(s^*, \gamma_m, d_{ver}) > 0$$

according to Lemma 3.10 such that

$$d_m(x, l_{s^*,-1,b}) = C_m d_{ver}(x, l_{s^*,-1,b}) \text{ for all } m \in \mathcal{M} \text{ and for all } b \in \mathbb{R}.$$

Defining the one-dimensional function $\bar{f}(b) = f(l_{s^*,-1,b})$ it follows that

$$\begin{aligned} \bar{f}(b) &= \sum_{m\in\mathcal{M}} d_m(Ex_m, l^*) \\ &= \sum_{m\in\mathcal{M}} C_m d_{ver}(Ex_m, l^*) \\ &= \sum_{m\in\mathcal{M}} C_m |a_{m2} - s^* a_{m1} + b| \end{aligned}$$

is a piecewise linear and convex function, since $C_m \geq 0$ for all $m \in \mathcal{M}$. Applying Lemma 1.9 we see that there exists a

$$b^0 \in median\{(a_{m2} - s^* a_{m1}, C_m)\}$$

minimizing $\bar{f}$ such that $l_{s^*,-1,b^0}$ contains at least one of the existing facilities and

$$f(l_{s^*,-1,b^0}) = \bar{f}(b^0) \leq \bar{f}(b^*) = f(l_{s^*,-1,b^*}),$$

completing the proof.

QED

A similar proof can establish that even in the case of mixed gauges there exists a median line passing through one of the existing facilities.

Lemma 5.7 *Let $\tilde{\gamma}_1, \tilde{\gamma}_2, \ldots, \tilde{\gamma}_M$ be gauges. Then for the line location problem $1l/\mathbb{R}^2/\cdot/\tilde{\gamma}_m/\sum$ there exists a median line passing through one of the existing facilities.*

Proof: We will only briefly sketch the idea of this proof. As in the proof of Lemma 5.6 we fix the slope of an optimal line $l_{s^*,-1,b^*}$ and derive constants C_m for all $m \in \mathcal{M}$ (using Lemma 5.1). As long as none of the existing facilities changes the side of l^*, this leads to the one-dimensional function $\bar{f}(b) = f(l_{s^*,-1,b})$ of the proof of Lemma 5.6. Then we can use the same technique as in the proof of Lemma 5.2 to derive the above result.

QED

Since even in the case of one identical metric for all existing facilities no structural result about median lines can be developed in general, this can neither be done in the case of mixed metrics.

Finally we discuss the property Cen1 for line location problems with mixed gauges.

Lemma 5.8 *Let $\tilde{\gamma}_1, \tilde{\gamma}_2, \ldots, \tilde{\gamma}_M$ be gauges. Then for the line location problem $1l/\mathbb{R}^2/\cdot/\tilde{\gamma}_m/\max$ there exists an optimal line which is at maximum distance from three of the existing facilities.*

Proof: We can reproduce the proof of Lemma 5.3 for which we define

$$A_m(g) = \{x : \tilde{\gamma}_m(Ex_m - x) \leq g\}.$$

Note that Corollary 4.3 does not use the fact that the sets A_m are translates of a common set A.

QED

Note that Cen1 is also true for line location problems with mixed, strictly monotone metrics, which can be shown by applying the proof of Lemma 5.3.

5.4 Algorithms for General Line Location Problems

In general, the computational problems that arise when calculating the distance between a point and a line are the same as those mentioned in Section 3.3. Ignoring these calculation difficulties as before we will present (later in this section) an algorithm that solves center line location problems for a large class of distance functions. First we mention that it seems to be more difficult to implement an algorithm for the median objective function for general distance measures. If we face a problem for which we know the existence of an optimal line passing through one of the existing facilities (as e.g. in the case of gauges or mixed norms) we can solve M one-dimensional optimization problems for all existing facilities $Ex_m \in \mathcal{E}x$. For lines passing through $Ex_m = (a_{m1}, a_{m2})$ our objective can be reduced to

$$\min_{s_1,s_2} \sum_{m\in\mathcal{M}} d(Ex_m, l_{s_1,s_2,-s_1a_{m1}-s_2a_{m2}})$$

$$s.t. \;\; s_1^2 + s_2^2 = 1.$$

This can be solved by line search on the circle, or by checking vertical lines separately. In the latter case we have to solve the following problem for all non-vertical lines (for all existing facilities $Ex_m = (a_{m1}, a_{m2})$):

$$\min_{s} \sum_{m\in\mathcal{M}} d(Ex_m, l_{s,-1,a_{m2}-sa_{m1}}).$$

The details of this approach are strongly dependent on the distance measure d.

In order to solve center problems with gauges, metrics, or even with mixed distances we can use Theorem 1.2 (on page 20) and proceed with the following steps.

Algorithm 7 (for $1l/\mathbb{R}^2/\cdot/d_m/\max$)

Input: Distances d_m, existing facilities Ex_m, weights $w_m > 0$ for all $m \in \mathcal{M}$ and a real number $\epsilon > 0$.

Output: Approximate objective value z of $1l/R^2/\cdot/d_m/\max$ with $|z - z^*| \le \epsilon$ and corresponding set of solutions $\mathcal{L}^*$.

1. Choose an arbitrary line l and define

$$\begin{aligned} z^{\max} &:= g(l), \\ z^{\min} &:= 0. \end{aligned}$$

2. $z := \frac{z^{\max}+z^{\min}}{2}$.

3. Determine

$$\begin{aligned} A_m(z) &:= conv(\{x : w_m d_m(Ex_m, x) \le z\}) \text{ for all } m \in \mathcal{M} \text{ and} \\ \mathcal{A}(z) &:= \{A_m(z) : m \in \mathcal{M}\}. \end{aligned}$$

4. Solve the line transversal problem for the family $\mathcal{A}(z)$. If there exists a line transversal l then

$$\begin{aligned} z^{\max} &:= z, \text{ else} \\ z^{\min} &:= z. \end{aligned}$$

5. If $|z^{\max} - z^{\min}| \ge \epsilon$ then goto Step 2.
Otherwise STOP.
Output: An approximate solution is $z^* := z$, and $\mathcal{L}^*$ is the set of all line transversals for $\mathcal{A}(z^*)$.

It is obvious that the performance of this algorithm depends mainly on Steps 3 and 4. In the case of norms or gauges, Step 3 is easy, since the same set B (the unit ball of the norm or gauge) simply has to be scaled and translated to determine the sets $A_m(z)$. On the other hand, for general metrics, the sets $A_m(z)$ may have to be computed from scratch in each iteration.
Since the line location problem is continuous, the binary search leads only to an approximate solution of the problem, and the worst-case running time is given by $(\lg(g(l)) - \lg(\epsilon))O(\text{Step 3} + \text{Step 4})$, where l is the starting line of Step 1. From transversal theory the following algorithms can be used to implement Step 4.

- In the general case $(1l/\mathbb{R}^2/\cdot/d_m/\max)$, i.e., if we allow any distance measure d_m for each of the existing facilities, the set $\mathcal{A}(z)$ consists of M convex sets $conv(A_m(z))$ for all $m \in \mathcal{M}$. Using the algorithm of [Her89], Step 4 can be carried out in $O(M \log M)$ time. This algorithm of Hershberger [Her89] has been shown to be optimal for line stabbing of convex sets, see [ARW89].

- Even in the case of translates (i.e., for the unweighted center problem $1l/\mathbb{R}^2/\cdot/w_m = 1/\tilde{\gamma}/\max$ with the same gauge for all existing facilities) the $O(M \log M)$ time complexity of [Her89] is best possible for Step 4 (see again [ARW89]). On the other hand, for this problem we can also use the line stabbing algorithm of [EW89], since the set $\mathcal{A}(z)$ consists of translates of a convex set. This algorithm runs in optimal $O(M \log M)$ time (like the algorithm of [Her89]), but if the sets in $\mathcal{A}(z)$ are pairwise disjoint its complexity is only linear.

- To solve the weighted problem with gauges $1l/\mathbb{R}^2/\cdot/\tilde{\gamma}/\max$ we use the optimal algorithm of [Her89] again, to find a stabbing line in Step 4 in time $O(M \log M)$.

- The same lower bound of $\Omega(M \log M)$ holds also for stabbing M line segments (i.e., solving Step 4 for a problem with different t-distances for all existing facilities, i.e., for $1l/\mathbb{R}^2/\cdot/(d_t)_m/\max$).

- If each of the distance measures d_m has a polygonal unit ball with G_m extreme points, then we can use another algorithm of [Her89] to implement Step 4 in $O(\sum_{m \in \mathcal{M}} G_m \log(\sum_{m \in \mathcal{M}} G_m))$ time. This algorithm is superior especially if the sets $A_m(z)$ are not translates of a common set A, since in this case faster algorithms are available (see the next remark).

- If we want to solve an unweighted problem with the same polyhedral gauge for all existing facilities, i.e., $1l/\mathbb{R}^2/w_m = 1/\tilde{\gamma}_B/\max$, then we can use the transversal algorithm for stabbing translates of a convex polygon, developed by [Ede85]. His algorithm runs in $O(GM)$ time. For the case that we have the same block norm for all existing facilities ($1l/\mathbb{R}^2/w_m = 1/\gamma_B/\max$) this yields that Step 4 can be performed in $O(GM)$ time. This is equal to the overall time complexity of Algorithm 4 which has been shown to be $O(GM)$ in Section 3.4. The latter exact algorithm is preferable in this case.

- Furthermore, we mention that [Ede85] also presented a linear time algorithm for stabbing M rectangles. Thus the time complexity of Step 4 can be improved, if all sets $A \in \mathcal{A}(z)$ are rectangles.

5.5 Summary

At the end of this chapter, Table 5.1 summarizes the properties discussed in this chapter and in Chapter 3. In this summary we additionally include the n-dimensional case of hyperplane location which will be discussed in Chapter 7. The properties Med1, Med2, Cen1, and Cen2 have been introduced in Section 1.3, see Table 1.6. Additionally, we write Med0 for the following property.

Med0 There exists a median line passing through one of the existing facilities.

If Med1 is true for a distance measure then obviously Med0 holds, but since this is the only difference between the case of strictly monotone metrics and mixed gauges, Med0 is included in Table 5.1.

Table 5.1. Properties of line location problems

	Med0	Med1	Med2	Cen1	Cen2
t-distances	yes	yes	yes	yes	yes
norms	yes	yes	yes	yes	yes
gauges	yes	no	no	yes	yes
strictly monotone metrics	no	no	no	yes	no
metrics	no	no	no	no	no
mixed gauges	yes	no	no	yes	no
norms in $\mathbb{R}^n$	yes	yes	yes	yes	no

Note that Cen1 is the only property which holds for all discussed distance measures, with the exception of arbitrary metrics. Furthermore we see that except for the case of t-distances and norms, no two rows of the table coincide.

6 RESTRICTED LINE LOCATION PROBLEMS

When applying location theory to model real-life problems one often has to take into account restrictions for the set of feasible solutions. Such additional constraints often change the whole structure of the model, and thus many theoretical results may be useless in practice. Therefore, many papers incorporate different types of restrictions into the models of classical location theory. One kind of restriction that is often used in continuous location of point-shaped facilities is the introduction of a **forbidden region** R (also called a **restricting set**) in the interior of which the new facilities cannot be located. Forbidden regions may represent lakes, natural habitats, industrial areas or already existing facilities, which are not allowed to be too close to the new facilities. For an overview on restricted classical location problems, the reader is referred e.g. to [Ham95, HN95].

For the case of line location, a given forbidden region R must not be intersected by the new line-shaped facility. The only problem of this type which has been studied in the literature so far is the location of a center hyperplane with respect to the Euclidean distance where the forbidden region is the convex hull of the existing facilities, see [Rob91]. In this chapter, however, we discuss arbitrary forbidden sets R through the interior of which the new line is not allowed to pass. The results of this chapter are the basis for [Sch98b].
As possible applications for restricted line location problems one can think of a highway that is not allowed to pass through a natural habitat, or a conveyor belt that cannot be within a dangerous area.

At the end of this chapter we will introduce some other types of restrictions to which the results of problems with restricting sets can be applied. Other types of restrictions may be for example **enforced regions** F, i.e., the new line must pass through a specified given set F. The case that the line is forced to pass through one given point, i.e., $F = \{x\}$, has been discussed in [MN83] for the Euclidean distance and in [KM93] for the location of hyperplanes with Euclidean distance.
Now suppose we have given an area in the plane (a restricting set R) that must not be intersected by any line. Then the **restricted line location problems** can be written as

$$\begin{array}{lcl} \min f(l) & \quad or \quad & \min g(l) \\ s.t. \;\; l \cap int(R) = \emptyset & & s.t. \;\; l \cap int(R) = \emptyset, \text{ respectively.} \end{array}$$

In the classification scheme these problems are denoted by $1l/\mathbb{R}^2/R/d/\sum$ and $1l/\mathbb{R}^2/R/d/\max$, respectively. In the following we will give an equivalent formulation of restricted line location problems using a set of forbidden lines

$$R_l := \{\text{lines } l : l \cap R \neq \emptyset\}$$

instead of the original set $R \subseteq \mathbb{R}^2$.

6.1 Restricted Problems with Vertical Distance

First we consider the problems $1l/\mathbb{R}^2/R/d_{ver}/\sum$ and $1l/\mathbb{R}^2/R/d_{ver}/\max$, where the restriction is given by a forbidden set $R \subseteq \mathbb{R}^2$. As discussed in Chapter 2 the objective functions of both problems are convex if no restrictions are given. Convexity fails to hold when we introduce a restricting set R. Nevertheless it is possible to derive structural results for the set of optimal lines $\mathcal{L}_R^*$ of the restricted problem using the dual interpretation of the problem which has been introduced in Section 2.2. To do this, we will proceed as follows. First we will transform restricted line location problems with vertical distance to the dual space. Then we show that in the dual space problems with a forbidden set can easily be solved using the theory developed in Section 1.5, which we then apply to the special case of $1l/\mathbb{R}^2/R/d_{ver}/\sum$ and $1l/\mathbb{R}^2/R/d_{ver}/\max$, respectively.

Recall that the lines

$$L_m = T(Ex_m) \text{ for } m \in \mathcal{M}$$

are the transformed existing facilities and that in the dual space the convex objective functions f and g can be written as

$$f(X) = \sum_{m \in \mathcal{M}} w_m d_{ver}(X, T(Ex_m)) \text{ and}$$

$$g(X) = \max_{m \in \mathcal{M}} w_m d_{ver}(X, T(Ex_m)), \text{ respectively.}$$

Now we have to transform the restricting set R to the dual space. To this end, we need the following definition.
Let $R \subseteq \mathbb{R}^2$ be a given set.

$$\begin{aligned} R_l &:= \{\text{lines } l : l \cap R \neq \emptyset\} \text{ and} \\ R^T &:= T(R_l) = \{X : T^{-1}(X) \cap R \neq \emptyset\}. \end{aligned} \tag{6.1}$$

Note that R^T is the set of all points in the dual space corresponding to lines which intersect the forbidden region R in the primal space. For this transformation of R to the dual space we have the following property which was already mentioned in [KA97] and [SA95] for a similar dual transformation.

Lemma 6.1 ([SA95]) *Let R be convex. Then $X \in \partial R^T$ if and only if the line $T^{-1}(X)$ touches R.*

With Lemma 6.1 we can conclude that

$$int(R^T) = \{X : T^{-1}(X) \cap int(R) \neq \emptyset\}$$

and hence the restricted line location problem in the primal space is equivalent to the following problem in the dual space:

$$\min h(X) \qquad\qquad s.t.\ \ X \notin int(R^T), h \in \{f, g\}.$$

Since both functions f and g are convex, this is a special case of (ROL) and we can use the theory developed in Section 1.5 to solve such problems. To this end we introduce a forbidden set R^D in the dual space and formulate the following restricted problems. Note that we use the denotation R^D for any restricting set in the dual problem, whereas the set R^T always is derived from a given set R in the primal space, according to 6.1.

$$\begin{array}{lcl} \min f(X) & \quad\text{or}\quad & \min g(X) \\ s.t.\ X \notin int(R^D) & & s.t.\ X \notin int(R^D). \end{array}$$

Both problems are special cases of (ROL). Since they have been derived from line location problems, they can be classified as $1l/\mathbb{R}^2/R^D/d_{ver}/\sum$ and as $1l/\mathbb{R}^2/R^D/d_{ver}/\max$, where the restriction R^D indicates that we have given a restricting set in the dual space. The next result follows immediately from Theorem 1.6 (see page 27).

Theorem 6.1 *Let R^D be a convex forbidden set in the dual space. Then there exists an optimal solution $X^*_{R^D}$ of (ROL) such that*

- *either $X^*_{R^D}$ is an optimal solution for the unrestricted problem (OL),*
- *or*

  for the median problem *:*

  $$X^*_{R^D} \in Cand = \{X : X \in \partial R^D \cap L_m \text{ for one } m \in \mathcal{M}\},$$

for the center problem :

$$X^*_{R^D} \in Cand = \{X : X \in \partial R^D \cap h \text{ for } h \in \mathcal{K}^{cen}_{weights}\}.$$

In both cases none of the one-dimensional intersections has to be considered.

Hence, in order to solve problems in the dual space with a convex forbidden set R^D we first determine the set $\mathcal{X}^*$ of all optimal solutions using Theorem 1.3. Then we only have to investigate the intersection points of all cell boundaries with the boundary of the restricting set R^D, yielding an efficient geometric approach to solve the restricted problem.

Unfortunately, in most cases we cannot use Theorem 6.1 if the restricting set R^D has been derived from a given restricting set R in the primal space, i.e.,rest $R^D = T(R_l) = R^T$. In particular, the next Lemma shows that for all sets R with non-empty interior the transformed set $T(R_l) = R^T$ is never convex if there exists any feasible line for the original problem.

Lemma 6.2 *Let $R \subseteq \mathbb{R}^2$. Then we have the following statements:*

1. *If R is connected, then $R^T = (conv(R))^T$.*
2. *R^T is convex if and only if $|R| = 1$ or if there exists is a vertical line contained in $conv(R)$ or if R is a vertical line segment.*

Proof:

1. This follows from the fact that a line l meets a set R if and only if l meets $conv(R)$, see Lemma 1.5.
2. If $R = \{x\}$ consists of only one point x then R^T is a (convex) line, and if all non-vertical lines intersect R then R^T is the whole dual space. If R is a vertical line segment with endpoints X and Y, then R^T is the (convex) strip between the parallel lines $T(X)$ and $T(Y)$.

   For the other direction, we distinguish the following two cases.

   Case 1 : Suppose that R is not contained in a vertical line, i.e., there exist two points $X = (x_1, x_2)$, $Y = (y_1, y_2) \in R$ with $x_1 < y_1$. Then we can choose a non-vertical line $l = l_{s,-1,b}$ not intersecting R and a point $P = (p_1, p_2) \in l$ such that $x_1 < p_1 < y_1$ and $P \notin aff(X, Y)$. Consider the two lines l_1 through X and P and l_2 through Y and P. For the slopes s_1 and s_2 of these lines we have $s_1 > 0$ and $s_2 < 0$. All three lines intersect in P, i.e., $T(l), T(l_1)$, and $T(l_2)$ are points on the same line $T(P)$ (see Lemma 2.3). Furthermore, $s_2 < s < s_1$ and thus $T(l)$ is a convex combination of $T(l_1)$ and $T(l_2)$. Since $T(l) \notin T(R_l)$, but $T(l_1), T(l_2) \in T(R_l)$, it follows that $R^T = T(R_l)$ is not convex.

   Case 2 : Suppose that R is contained in a vertical line. Since R is not a line segment, we can find two points $X_1 = (x, b_1), X_2 = (x, b_2) \in R$

and some point $Y = (x, b) \notin R$ between X_1 and X_2, i.e., $b_1 < b < b_2$ without loss of generality. Thus the horizontal line through Y is a convex combination of the horizontal lines through X_1 and X_2, but it does not intersect R.

QED

Lemma 6.2 shows that a simple enumeration of all lines in the candidate set *Cand* defined in Theorem 6.1 is not sufficient in order to solve the restricted line location problem. But we can derive the following result.

Theorem 6.2 *Let $R \subseteq \mathbb{R}^2$ be a forbidden set in the plane. For the restricted line location problems $1l/\mathbb{R}^2/R/d_{ver}/\sum$ and $1l/\mathbb{R}^2/R/d_{ver}/\max$ the following holds: If no optimal line for the unrestricted problem is feasible for the restricted problem then every line solving the restricted line location problem is a tangent to R.*

Proof: In the dual space we conclude from Theorem 1.4 that all optimal solutions lie on the boundary of R^T. If R is convex, we can directly apply Lemma 6.1 to get the result. If R is not convex, we consider $conv(R)$ according to Lemma 6.2 and obtain an optimal solution which is a tangent to $conv(R)$ and therefore also to R (see Lemma 1.5 on page 19).

QED

If R is an arbitrary set, there may exist infinitely many tangents which have to be considered to solve the restricted problem. For polygonal sets, however, there exists a finite candidate set for the optimal solution of the restricted problem. To prove this we need the following lemma, which is illustrated in Figure 6.1.

Lemma 6.3 *Let R be a simple polygon. Then R^T is a non-convex (non-finite) polygon in the dual space and the following statements hold:*

1. *(s, b) is a vertex of R^T if and only if $l_{s,b}$ contains a non-vertical facet of $conv(R)$.*

2. *v is a vertex of $conv(R)$ if and only if $T(v)$ contains a non-vertical facet of R^T.*

Proof: According to Lemma 6.1 we only have to check the tangents of R. Exactly those dual points that correspond to tangents passing through a vertex v of R lie on the line $T(v)$ (see Lemma 2.3) and therefore form an edge of R^T. Since each facet of R contains two vertices, we conclude that exactly those lines l containing an edge of $conv(R)$ correspond to points on two edges of R^T, i.e., to the vertices of R^T.

QED

Figure 6.2 gives an illustration of Theorem 6.3.

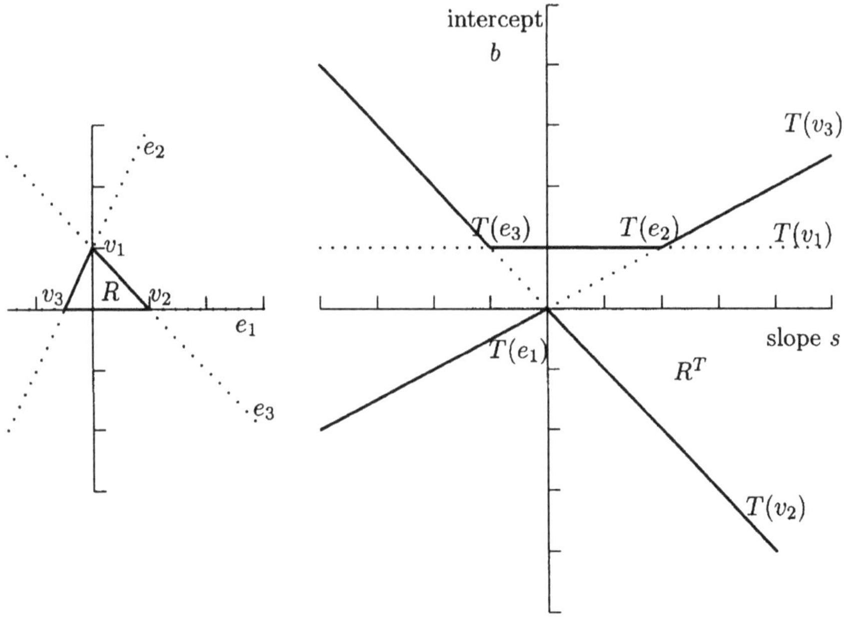

Figure 6.1. Transformation of a triangle R to the dual space.

Theorem 6.3 *Let R be a simple polygon.*

1. *For the restricted problem $1l/\mathbb{R}^2/R = Polygon/d_{ver}/\sum$ we have:*
   - *Either an optimal line for the unrestricted problem is feasible*
   - *or there exists an optimal line for the restricted problem which contains a facet of R or which passes through one of the existing facilities and a vertex of R, i.e., there exists a line $l \in Cand_{Poly}$ where*

$$\begin{aligned} Cand_{Poly} &:= \{lines\ l : l \text{ contains a facet of } R\} \\ &\cup \{lines\ l : \text{ there exist } m \in \mathcal{M}, v \in Ext(R) \text{ with } \\ &\quad Ex_m, v \in l\}. \end{aligned}$$

2. *For the restricted problem $1l/\mathbb{R}^2/R = Polygon, w_m = 1/d_{ver}/\max$ we have:*
   - *Either an optimal line for the unrestricted problem is feasible*
   - *or there exists an optimal line for the restricted problem which contains a facet of R or which passes through a vertex of R and is at maximum distance from two of the existing facilities, i.e., there exists a line $l \in Cand_{Poly}$ where*

$$\begin{aligned} Cand_{Poly} &:= \{lines\ l : l \text{ contains a facet of } R\} \\ &\cup \{lines\ l : \text{ there exist } m_1, m_2 \in \mathcal{M}, v \in Ext(R) \text{ with } \\ &\quad v \in l \text{ and } g(l) = w_{m1} d_{ver}(Ex_{m_1}, l) = w_{m_2} d_{ver}(Ex_{m_2}, l).\} \end{aligned}$$

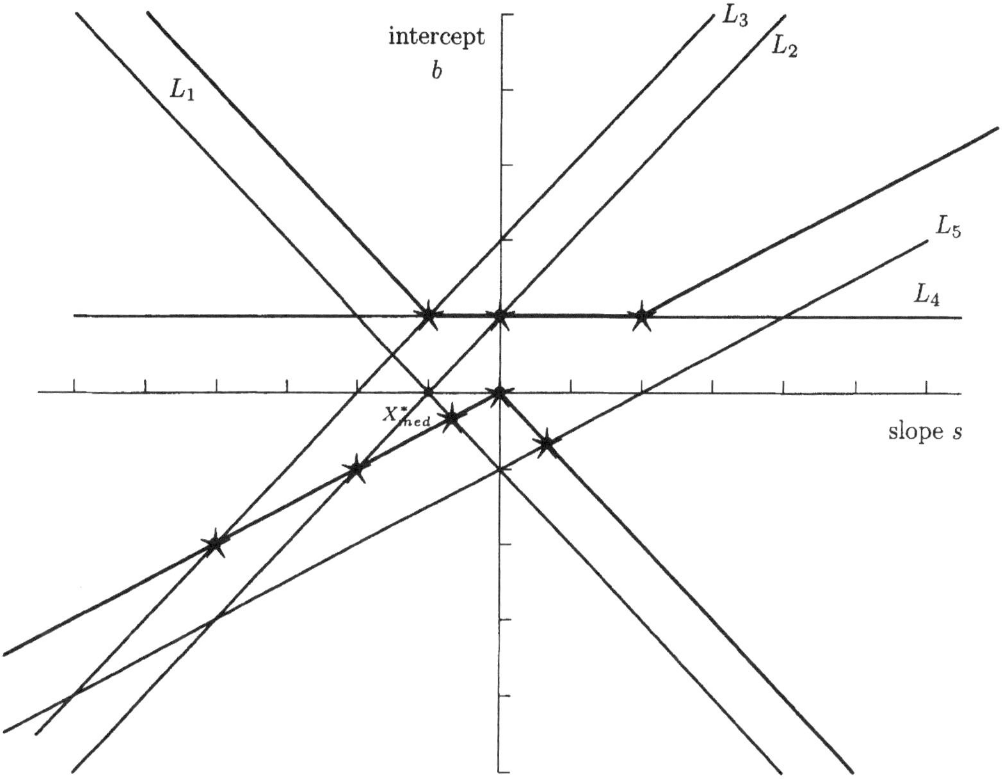

Figure 6.2. The restricting set in the dual space. The candidate points for the median problem are marked by stars.

Proof: Using Lemma 6.2 we transform $conv(R)$ to the dual space and apply Corollary 1.1 (see page 31).

Median problem : We know that there exists an optimal solution X, which is

- either an intersection point between a construction line $T(Ex_m)$ and ∂R^T; in this case the line $T^{-1}(X)$ touches ∂R (see Lemma 6.1) and contains Ex_m (see Lemma 2.3),
- or a reflexive vertex of R^T; in this case the line $T^{-1}(X)$ contains a facet of $conv(R)$ (see Lemma 6.3).

Center problem : Analogously we know that there exists an optimal solution X, which is

- either an intersection point between a construction line $l \in \mathcal{K}^{cen}_{weights}$ and ∂R^T; in this case the line $T^{-1}(X)$ touches ∂R (see Lemma 6.1) and is at maximum distance from two of the existing facilities.

- or a reflexive vertex of R^T; in this case the line $T^{-1}(X)$ contains a facet of $conv(R)$ (see Lemma 6.3).

QED

Note that in the case of vertical distances d_{ver} it is also possible to calculate the set of all optimal solutions of the restricted problem $\mathcal{X}_R^*$ using the following formula. If $t_R^* = f(l_R^*)$ denotes the objective value of the restricted problem then the set of optimal solutions in the dual space is given by the intersection of the level set $L_{\leq}(t_R^*)$ and the boundary of the transformed restricting set, i.e., $\mathcal{X}_R^* = L_{\leq}(t_R^*) \cap \partial R^T$, which corresponds to a set of tangents in the primal space.

6.2 Generalization to Block Norms

The main advantage of the vertical distance is, that the unrestricted line location problems are convex. This property does not transfer to block norm distances, even for l_1 the (unrestricted) problem becomes non-convex. For block norm distances, however, the separation argument of Section 3.4 helps again to solve the problem. As in the unrestricted case we can solve restricted line location problems with block norm distances by solving the problem for all fundamental directions $b_1, b_2 \ldots b_G$ individually (by transforming these problems to d_{ver}) and then taking the best of the obtained solutions. If the restriction is a simple polygon R we can therefore generalize the results of Theorem 6.3 to block norm distances d_B.

Theorem 6.4 *Let R be a simple polygon. For the restricted problems $1l/\mathbb{R}^2/R =$ Polygon$/d_B/\sum$ and $1l/\mathbb{R}^2/R =$ Polygon$/d_B/\max$ we have:*

- *Either an optimal line for one of the corresponding unrestricted problems $1l/\mathbb{R}^2/\cdot/d_{b_g}/\cdot$, $g = 1, \ldots, G$, is feasible and optimal for the restricted problem or*

- **for the median problem:** *there exists an optimal line for the restricted problem which contains a facet of R or which passes through one of the existing facilities and a vertex of R.*

  for the center problem: *there exists an optimal line for the restricted problem which contains a facet of R or which is at maximum distance from two of the existing facilities and passes through a vertex of R.*

One observation shall be emphasized here. It may happen that an optimal line for the restricted problem is neither optimal for the unrestricted problem nor tangent to the restricting set R. Hence Theorem 6.2 does not hold for block norm distances in general. This will be demonstrated in the following example.

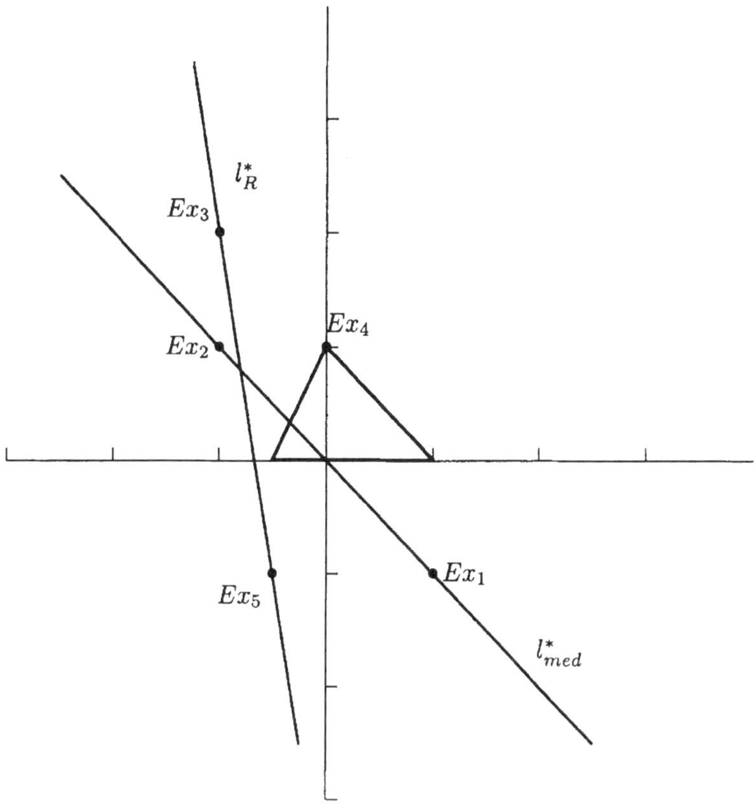

Figure 6.3. An optimal solution l^*_R of a restricted problem (with block norm distance) which is neither optimal for the unrestricted case nor tangent.

Example 6.1 *We use the set of existing facilities depicted in Figures 2.1 (page 38) and 6.1. In this example the distance function d_γ is derived from the block norm*

$$\gamma(x) = \frac{3}{2}|x_1| + |x_2|, x = (x_1, x_2)$$

with extreme points $b_1 = (0,1)$ and $b_2 = (\frac{3}{2}, 0)$. Then the optimal solution l^* of the unrestricted problem $1l/\mathbb{R}^2/\cdot/d_\gamma/\sum$ equals the optimal solution of the analogous problem with vertical distance $d_{ver} = d_{b_1}$. With the triangle introduced in Figure 6.1 as restriction, l^* is forbidden and one optimal solution l^*_R (also minimizing d_{hor}) is shown in Figure 6.3. This line is neither optimal for the unrestricted case, nor tangent to the restricting set R, i.e. Theorem 6.2 does not hold in this example.

In general, it is necessary to determine all optimal solutions of the unrestricted problems for each fundamental direction $d_{b_1}, \ldots, d_{b_G}$ (applying Corollary 4.4 on page 92) to check if any of these optimal solutions is also feasible in the restricted case. As a consequence it is easy to determine the complete set of optimizers for $1l/\mathbb{R}^2/R = Polygon/d_b/\cdot$. That this can be relaxed if we are only interested in *one* optimal solution will be shown in the next section.

6.3 Generalization to Arbitrary Norms

The extension of the results of the previous sections to arbitrary norms causes the following difficulty. Given a norm γ we know that any point t on the boundary of the dilated unit ball

$$\lambda B_\gamma = \{\lambda t : \gamma(t) = 1\}$$

may touch the optimal line. Hence the only straightforward conclusion is that the optimal line for the restricted problem either is an optimal solution of one of the (infinitely many) corresponding problems with distance d_t ($1l/\mathbb{R}^2/\cdot/d_t/\cdot$) for all $t \in \mathbb{R}^2$, or the optimal line is tangent to the restricting set R. Algorithmically this is certainly not very helpful, but using the following lemma (for the vertical distance d_{ver}) it is possible to determine a finite candidate set in the case of arbitrary norms, too.

Lemma 6.4 *Let R be a polygonal set and let $\mathcal{L}^*$ be the set of optimal lines for the unrestricted problem with vertical distance ($1l/\mathbb{R}^2/\cdot/d_{ver}/\sum$ or $1l/\mathbb{R}^2/\cdot/d_{ver}/\max$). Moreover, suppose that there are lines in $\mathcal{L}^*$ which intersect $int(R)$ and lines in $\mathcal{L}^*$ which are feasible for the restricted problem. Then there exists a line $l \in \mathcal{L}^*$ which is*

for the median problem *feasible for the restricted problem and which is tangent to R and passes through one of the existing facilities.*

for the center problem *feasible for the restricted problem and which is tangent to R and is at maximum distance from two of the existing facilities.*

Proof: Define

$$\begin{aligned} R_l &:= \{\text{lines } l : l \cap R \neq \emptyset\} \text{ and} \\ R^T &:= T(R_l) = \{X : T^{-1}(X) \cap R \neq \emptyset\} \end{aligned}$$

according to page 117. Then both sets

$$T(\mathcal{L}^*) \cap int(R^T) \text{ and } T(\mathcal{L}^*) \cap \mathbb{R}^2 \setminus int(R^T)$$

are non-empty sets in the dual space. Thus the boundaries of $T(\mathcal{L}^*)$ and R^T intersect in a common point X, corresponding to a line $l = T^{-1}(X)$, (using the structure of R^T described in Lemma 6.3 and because Lemma 4.5 shows that $T(\mathcal{L}^*)$ is connected). Since

$$X \in \partial R^T,$$

the line l is tangent to R according to Theorem 6.2. Furthermore, every point $Y \in \partial T(\mathcal{L}^*)$ is contained in at least one construction line such that for the median problem there exists an $m \in \mathcal{M}$ with $X \in T(Ex_m)$. Thus, according to Lemma 2.3, $Ex_m \in l$ in the case of the median problem. For the center problem l is at maximum distance from two of the existing facilities.

QED

Theorem 6.5 *Let R be a simple polygon. For the restricted problems $1l/\mathbb{R}^2/R =$ Polygon$/\gamma/\cdot$ there exists an optimal line which*

for the median problem :

- *contains a facet of R or*
- *passes through one of the existing facilities and a vertex of R or*
- *passes through two of the existing facilities.*

for the center problem :

- *contains a facet of R or*
- *is at maximum distance from two of the existing facilities and passes through a vertex of R or*
- *is at maximum distance from three of the existing facilities.*

Proof: We prove this result for the median objective function. (For the center function the proof can be provided analogously.) Let d be the distance derived from the norm γ. Now take an optimal line l^* for the restricted problem $1l/\mathbb{R}^2/R =$ *Polygon*$/\gamma/\sum$ that does not satisfy any of the above properties. Choose $t \in \mathbb{R}^2$ such that

$$d(x, l^*) = d_t(x, l^*) \text{ for all } x \in \mathbb{R}^2$$

according to Lemma 3.12. Consider the corresponding location problem with distance d_t instead of d, i.e., $1l/\mathbb{R}^2/R =$ *Polygon*$/d_t/\sum$. Let $\mathcal{L}_t^*$ denote the set of optimal solutions for the unrestricted problem with distance d_t, (i.e., for $1l/\mathbb{R}^2/\cdot/d_t/\sum$). By considering the following three cases we construct another optimal line l^0 that satisfies one of the required properties. Note that Theorem 6.3 and Lemma 6.4 can be applied not only to d_{ver}, but also to d_t by rotation.

- If all lines $l \in \mathcal{L}_t^*$ intersect $int(R)$, we know from Theorem 6.3 that there exists a line l^0 which is optimal for the restricted problem $1l/\mathbb{R}^2/R =$ *Polygon*$/d_t/\sum$ and passes through an existing facility and a vertex of R or which contains a facet of R.
- If no line $l \in \mathcal{L}_t^*$ intersects $int(R)$, then all optimal lines for the unrestricted problem are also feasible in the restricted case. According to Theorem 3.1

(the weak incidence property for the distance d_t) there exists an optimal line l^0 passing through two of the existing facilities.

- If there exists a line in $\mathcal{L}_t^*$ which intersects $int(R)$ and there also exists a line in $\mathcal{L}_t^*$ which does not intersect $int(R)$ we conclude from Lemma 6.4 that there exists a feasible line $l^0 \in \mathcal{L}_t^*$ for the restricted problem which passes through one of the existing facilities and a vertex of R.

For the new line l^0 we can conclude from Lemma 3.11 that

$$d(x, l^0) = \min_{t' \in \mathbb{R}^2} d_{t'}(x, l^0) \leq d_t(x, l^0)$$

for all $x \in \mathbb{R}^2$. Summarizing, we estimate the objective value of l^0.

$$\begin{aligned} f(l^*) &= \sum_{m \in \mathcal{M}} w_m d(Ex_m, l^*) \\ &= \sum_{m \in \mathcal{M}} w_m d_t(Ex_m, l^*) \\ &\geq \sum_{m \in \mathcal{M}} w_m d_t(Ex_m, l^0) \\ &\geq \sum_{m \in \mathcal{M}} w_m d(Ex_m, l^0) = f(l^0), \end{aligned}$$

and thus l^0 is also optimal for the restricted problem and satisfies at least one of the above properties.

QED

We conclude this section by suggesting two algorithms that find an optimal line for restricted line location problems with median and center objective function, respectively. In both algorithms we allow all distance measures d derived from norms. First we give the algorithm for the median problem.

Algorithm 8 (for $1l/\mathbb{R}^2/R = Polygon/\gamma/\sum$)

Input: Distance d, derived from a norm γ, $Ex_m, w_m > 0$ for all $m \in \mathcal{M}$ and forbidden polygonal set R

Output: A line l^* solving $1l/\mathbb{R}^2/R = Polygon/\gamma/\sum$

1. $z^* := \infty$.

2. $R := conv(R)$.

3. Determine the set of candidate lines

$$\begin{aligned} Cand &:= \{l : l \text{ contains a facet of } R\} \\ &\cup \{l : \text{ there exist } m \in \mathcal{M}, v \in Ext(R) \text{ with } Ex_m, v \in l\} \\ &\cup \{l : \text{ there exist } k, l \in \mathcal{M} \text{ with } Ex_k, Ex_l \in l\}. \end{aligned}$$

4. For all lines $l \in Cand$ do

1. Determine $f(l)$.
2. If $f(l) < z^*$ then $z^* := f(l)$ and $l^* := l$.

5. Output: l^* with objective value z^*.

For the center line problem with a restricting polygonal set R we need to determine all lines which are at maximum distance from two of the existing facilities while passing through a vertex of R. To this end, we can use the points P_{jk} $(j, k \in \mathcal{M})$ which have already been needed in Algorithm 2 and have been introduced on page 68. Lemma 3.13 then implies that any line which is at maximum distance from Ex_k and from Ex_j while separating these two existing facilities contains P_{jk}. But in our case an optimal line l for the restricted problem may *not* separate the points Ex_k and Ex_j. Hence we additionally need to investigate all lines l parallel to $Ex_k - Ex_j$ if $w_k = w_j$. In the case that $w_k < w_j$ we define

$$P'_{kl} = \frac{w_k}{w_k - w_j} Ex_k - \frac{w_j}{w_k - w_j} Ex_j$$

and then show the following result analogously to Lemma 3.13.

Lemma 6.5 *Let $w_k d(Ex_k, l) = w_j d(Ex_j, l)$ for a distance d derived from a norm and a line l.*

1. *If $w_k < w_j$ then either $P_{kj} \in l$ or $P'_{kj} \in l$.*
2. *If $w_k = w_j$ then either $P_{kj} \in l$ or l is parallel to $Ex_k - Ex_j$.*

Now we are ready to formulate Algorithm 9.

Algorithm 9 (for $1l/\mathbb{R}^2/R = Polygon/\gamma/\max$)

Input: Distance d, derived from a norm γ, $Ex_m, w_m > 0$ for all $m \in \mathcal{M}$ and forbidden polygonal set R

Output: A line l^* solving $1l/\mathbb{R}^2/R = Polygon/\gamma/\max$

1. $z^* := \infty$.

2. $R := conv(R)$.

3. For all $j, k \in \mathcal{M}, j \neq k$, determine P_{jk}

4. Determine the set of candidate lines *Cand* consisting of

- $\{l : l$ contains a facet of $R\}$
- $\{l :$ there exist $j, k \in \mathcal{M}, v \in Ext(R)$ with $v, P_{jk} \in l\}$
- $\{l :$ there exist $j, k \in \mathcal{M}, w_j < w_k,$ and $v \in Ext(R)$ with $v, P'_{jk} \in l\}$
- $\{l :$ there exist $j, k \in \mathcal{M}, w_j = w_k,$ and $v \in Ext(R)$ with $v \in l$ and l is parallel to $Ex_j - Ex_k\}$
- $\{l :$ there exist $j, k, m \in \mathcal{M}$ with $P_{jk}, P_{jm} \in l\}$.

5. For all lines $l \in Cand$ do

1. Determine $g(l)$.
2. If $g(l) < z^*$ then $z^* := g(l)$ and $l^* := l$.

6. Output: l^* with objective value z^*.

6.4 Locating a Line Segment with Vertical Distance

In this section we discuss the location of line segments in the plane: Given a real number $l_0 > 0$ and a set $\mathcal{E}x = \{Ex_1, Ex_2, \dots, Ex_m\}$ of points in the plane representing the existing facilities, we are looking for a line segment S with Euclidean length l_0 such that the sum of vertical distances (or the maximum vertical distance) between the existing facilities and the line segment is minimized. Since the length l_0 of a line segment can be assumed to be proportional to its costs, the above constraint (that the Euclidean length of the segment equals l_0) allows only to consider those line segments that do not exceed a given budget. This problem is a generalization of the line location problem, but only very few literature about the location of line segments is available. For the Euclidean distance and the center objective function the location of a line segment has been studied in [ILY92, AEST93] and in [Pet97].

An implementation of a heuristic approach to find line segments minimizing the Euclidean distance has been given by [Erk96].

Formally, the problem of locating a line segment in the plane can be given by replacing *New* in the general formulation on page 9 by a line segment S; this formulation will now be repeated. Given a set of M existing facilities $\mathcal{E}x \subseteq \mathbb{R}^2$ with positive weights, and, additionally, a real number $l_0 \in \mathbb{R}$, find a line segment S with Euclidean length l_0. The objective is to minimize

$$f(S) = \sum_{m \in \mathcal{M}} w_m d_{ver}(Ex_m, S), \text{ or}$$

$$g(S) = \max_{m \in \mathcal{M}} w_m d_{ver}(Ex_m, S), \text{ respectively.}$$

A segment S^* minimizing f will be called a **median line segment**, and if it minimizes g we will refer to it as a **center line segment**. In the classification scheme these problems are denoted by $1S/\mathbb{R}^2/\cdot/d_{ver}/\sum$, and $1S/\mathbb{R}^2/\cdot/d_{ver}/\max$. To describe a line segment with a given length, three variables are necessary. Let l^S be the line passing through the line segment S. Then we define

$$S(s, b, x^-) = \{(x_1, x_2) : x_2 = sx_1 + b, x^- \leq x_1 \leq x^+\}$$

$$\text{where } x^+ := x^- + \frac{l_0}{\sqrt{1+s^2}}.$$

In this definition, the line segment $S = S(s, b, x^-)$ is described by the slope s and the intercept b of the line l^S passing through S, and by the smallest x-value x^- for which a point on the line l^S also lies on the segment S.
With this notation the vertical distance between a point $x = (x_1, x_2)$ and a line segment $S = S(s, b, x^-)$ is given by the following formula:

$$d_{ver}(x, S) = \begin{cases} |x_1 s - x_2 + b| & \text{if } x^- \leq x_1 \leq x^+ \\ \infty & \text{otherwise.} \end{cases}$$

Note that this definition of a line segment excludes the case of a vertical segment. Nevertheless this is no loss of generality since, if not all existing facilities have the same x_1-coordinate, all vertical line segments S have objective value $f(S) = g(S) = \infty$ and need not be considered.

One additional definition is needed, before we can discuss the location of line segments:

Definition 25 *Given a set of existing facilities $\mathcal{E}x$, the* **span** *of $\mathcal{E}x$ is given by*

$$span(\mathcal{E}x) = \max_{m \in \mathcal{M}} a_{m1} - \min_{m \in \mathcal{M}} a_{m1}.$$

Obviously, if $l_0 < span(\mathcal{E}x)$ then

$$f(S) = g(S) = \infty$$

for all line segments S with length l_0. Furthermore, we can state the following lemma.

Lemma 6.6 *For any line segment $S = S_{s,b,x^-}$ with length l_0 the following statements are equivalent:*

1. $f(S) < \infty$
2. $f(S) = f(l^S)$
3. $g(S) < \infty$
4. $g(S) = g(l^S)$
5. *s and x^- satisfy*

$$|s| \leq \sqrt{\left(\frac{l_0}{span(\mathcal{E}x)}\right)^2 - 1} \quad and$$

$$\max_{m\in\mathcal{M}} a_{m1} - \frac{l_0}{\sqrt{1+s^2}} \leq x^- \leq \min_{m\in\mathcal{M}} a_{m1}.$$

Proof:

1. ⇔ 2.

$$\begin{aligned} f(S) < \infty &\Leftrightarrow d_{ver}(Ex_m, S) < \infty \ \forall m \in \mathcal{M} \\ &\Leftrightarrow d_{ver}(Ex_m, S) = |a_{m1}s - a_{m2} + b| = d_{ver}(Ex_m, l^S) \\ &\Leftrightarrow f(S) = f(l^S). \end{aligned}$$

1. ⇔ 5.

$$\begin{aligned} f(S) < \infty &\Leftrightarrow d_{ver}(Ex_m, S) < \infty \ \forall m \in \mathcal{M} \\ &\Leftrightarrow x^- \leq a_{m1} \leq x^+ \ \forall m \in \mathcal{M} \\ &\Leftrightarrow \min_{m\in\mathcal{M}} a_{m1} \geq x^- \text{ and } \max_{m\in\mathcal{M}} a_{m1} \leq x^- + \frac{l_0}{\sqrt{1+s^2}} \\ &\Leftrightarrow \max_{m\in\mathcal{M}} a_{m1} - \frac{l_0}{\sqrt{1+s^2}} \leq x^- \leq \min_{m\in\mathcal{M}} a_{m1} \text{ and } \frac{l_0}{\sqrt{1+s^2}} \geq span(\mathcal{E}x) \\ &\Leftrightarrow \max_{m\in\mathcal{M}} a_{m1} - \frac{l_0}{\sqrt{1+s^2}} \leq x^- \\ &\qquad\qquad \leq \min_{m\in\mathcal{M}} a_{m1} \text{ and } |s| \leq \sqrt{\left(\frac{l_0}{span(\mathcal{E}x)}\right)^2 - 1}. \end{aligned}$$

1. ⇔ 3. and 2. ⇔ 4. Using that

$$\begin{aligned} \sum_{m\in\mathcal{M}} w_m d_{ver}(Ex_m, S) < \infty &\iff w_m d_{ver}(Ex_m, S) < \infty \text{ for all } m \in \mathcal{M} \\ &\iff \max_{m\in\mathcal{M}} w_m d_{ver}(Ex_m, S) < \infty, \end{aligned}$$

these equivalences can be derived directly.

QED

Corollary 6.1 repeats the results of Lemma 6.6 which will be used in the following, because they show that the line segment problem with vertical distance can be reduced to a line location problem.

Corollary 6.1 *Let S be a line segment and l^S be the line containing S.*

1. *If $f(S) < \infty$, then $|s| \le \sqrt{\left(\frac{l_0}{span(\mathcal{E}x)}\right)^2 - 1}$ and $f(S) = f(l^S)$.*

2. *If $g(S) < \infty$, then $|s| \le \sqrt{\left(\frac{l_0}{span(\mathcal{E}x)}\right)^2 - 1}$ and $g(S) = g(l^S)$.*

In order to find an optimal line segment, we therefore can proceed as follows: In a first step, find a line minimizing the sum of vertical distances to the existing facilities, such that the slope s of the line satisfies

$$|s| \le \sqrt{\left(\frac{l_0}{span(\mathcal{E}x)}\right)^2 - 1}.$$

In a second step, find the best line segment with length l_0 on that line.

The second step is simple, as the following shows: Suppose a line l is given and we want to locate the line segment S on the line l. Thus we are looking for x^- such that

$$f(S_{s,b,x^-}), \text{ or } g(S_{s,b,x^-})$$

is minimized, where s and b are defined by the line l. Then, according to Lemma 6.6, an optimal choice for x^- is

$$x^- := \min_{m \in \mathcal{M}} a_{m1}.$$

The main advantage of the first step is that we are looking for a line instead of a line segment. This problem can be classified as $1l/\mathbb{R}^2/R^D/d_{ver}/\sum$ with the restriction R^D indicating that only lines $l_{s,b}$ with $|s| \le a$ for a given number a are feasible. To solve this restricted line location problem we can use the following theorem.

Theorem 6.6 *Let S^* be the set of slopes belonging to optimal lines for the (unrestricted) line location problem $1l/\mathbb{R}^2/\cdot/d_{ver}/\sum$ or $1l/\mathbb{R}^2/\cdot/d_{ver}/\max$. Let the restriction $|s| \le a$ for the slope of the line be given. Then for all optimal lines $l^R = l_{s^R,b^R}$ of the restricted problem it holds that*

$$s^R \in S^* \cup \{a, -a\}.$$

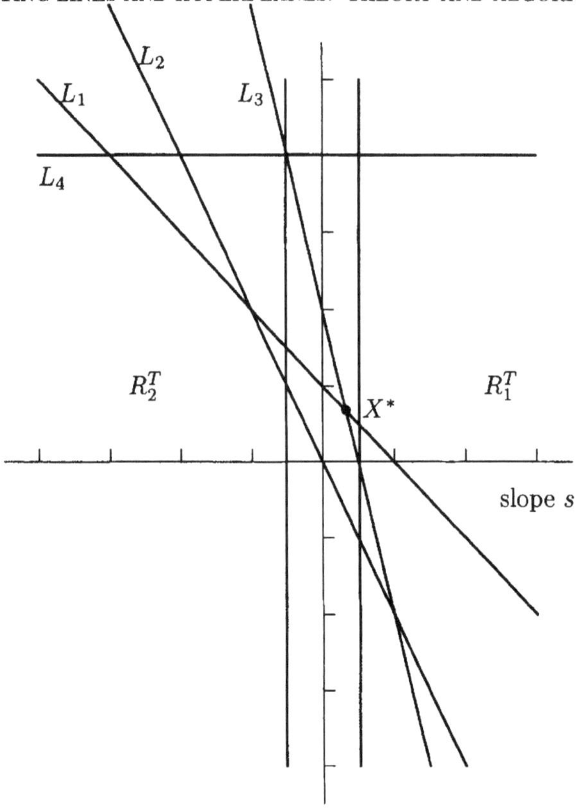

Figure 6.4. Locating a line segment: An example in the dual space.

Proof: We transform the problem to the dual space and consider

$$\min h(s,b)$$

$$s.t. \quad s \leq |a|$$

where $h \in \{f,g\}$, i.e., we have two convex restricting sets (for an example see Figure 6.4)

$$\begin{aligned} R_1^D &= \{X=(s,b) : s \geq a\} \text{ and} \\ R_2^D &= \{X=(s,b) : s \leq -a\}. \end{aligned}$$

If all optimal solutions $\mathcal{X}^*$ of the unrestricted problem are infeasible, then the convex set $\mathcal{X}^*$ is contained in exactly one of the sets R_i^D, $i \in \{1,2\}$. With Theorem 6.1 (see page 117) we conclude that there exists an optimal solution

$$X_{R^D}^* \in \partial R_i^D \cap \mathcal{K},$$

where $\mathcal{K} = \mathcal{K}^{med}$ in the median case and $\mathcal{K} = \mathcal{K}^{cen}_{weights}$ in the center case. Furthermore,

$$\begin{aligned} \partial R_1^D &= \{X = (s,b) : s = a\}, \text{whereas} \\ \partial R_2^D &= \{X = (s,b) : s = -a\}. \end{aligned}$$

Now we consider the line

$$l = T^{-1}(X^*_{R^D}).$$

Then the absolute value of the slope $|s|$ of l is equal to a. Furthermore, for the median problem it follows that l passes through one of the existing facilities, and for the center problem we analogously get that l is at maximum distance from two of the existing facilities.

QED

Note that it is also possible to prove the above result without the dual interpretation of the problem by using the convexity of the corresponding line location problem with vertical distance. This was done by the author in [Sch97].
To solve the original problem, i.e., to locate a line segment with length $l_0 > 0$, we finally can conclude:

Theorem 6.7 *For both problems* $1S/\mathbb{R}^2/\cdot/d_{ver}/\sum$ *and* $1S/\mathbb{R}^2/\cdot/d_{ver}/\max$ *there exist optimal line segments* S *with given length* l_0 *such that either*

- *S is contained in an optimal line of the corresponding line location problem* $1l/\mathbb{R}^2/\cdot/d_{ver}/\sum$ *and* $1l/\mathbb{R}^2/\cdot/d_{ver}/\max$*, respectively, or*
- *S is contained in an optimal line with fixed slope* $\sqrt{\left(\frac{l_0}{span(\mathcal{E}x)}\right)^2 - 1}$*, or*
- *S is contained in an optimal line with fixed slope* $-\sqrt{\left(\frac{l_0}{span(\mathcal{E}x)}\right)^2 - 1}$*.*

These results lead to the following algorithm for solving the line segment problem with respect to the vertical distance. Since the formulation of this algorithm is nearly the same for the median and for the center objective, we give it only once for both line segment problems.

Algorithm 10 (for solving $1S/\mathbb{R}^2/\cdot/d_{ver}/\sum$ and $1S/\mathbb{R}^2/\cdot/d_{ver}/\max$)

Input : $\mathcal{M}$, $\mathcal{E}x$, $w_m > 0$ for all $m \in \mathcal{M}$, l_0

Output : Line segment S_{s,b,x^-} with length l_0 solving $1S/\mathbb{R}^2/\cdot/d_{ver}/\sum$ or $1S/\mathbb{R}^2/\cdot/d_{ver}/\max$, respectively.

1. Calculate *span*. If $l_0 < span$ then
Output: All line segments S have infinite objective value.

2. Solve the corresponding line location problem. Let $\mathcal{L}^*$ be the set of optimal lines.

3. Let $s_0 := \sqrt{\left(\frac{l_0}{span}\right)^2 - 1}$

4. If there is an optimal line $l^* = l_{s^*,b^*} \in \mathcal{L}^*$ with slope $|s^*| \leq s_0$ then choose $s := s^*$, $b := b^*$, and $x^- := \min_{m \in \mathcal{M}} a_{m1}$.
Output: S_{s,b,x^-} is an optimal segment.

5. If all optimal lines $l^* \in \mathcal{L}^*$ have slopes s with $|s| > s_0$, then take any s with $l_{s,b} \in \mathcal{L}^*$ and choose $s := \text{sgn}(s)s_0$.

In the median case choose $b \in median\{(a_{m2} - a_{m1}s, w_m) : m \in \mathcal{M}\}$

In the center case set $b := \text{argmin}_{b' \in \mathbb{R}} \max_{m \in \mathcal{M}} w_m|a_{m2} - sa_{m1} + b'|$

and in both cases choose $x^- := \min_{m \in \mathcal{M}} a_{m1}$.
Output: S_{s,b,x^-} is an optimal line segment.

Note that the optimal segment is not unique in general, but s and x^- of the optimal segment are unique in Step 5. The complexity of Algorithm 10 applied to the median problem is $O(M)$, i.e., linear in the number of the existing facilities, since this is the best known complexity of Step 2 (see [Zem84]) and since the median in Step 5 can also be determined in linear time (see [BFP+72]). For the center problem we get a worst-case complexity of $O(M)$ since Step 2 and Step 5 can be solved by linear programming in linear time with the techniques of [Meg84].

Corollary 6.2 *There exists an optimal line segment S such that one existing facility Ex_m is contained in S.*
In particular: There exists an optimal line segment S such that two of the existing facilities are contained in S or one existing facility is contained in S and in this case $x^- = \min_{m \in \mathcal{M}} a_{m1}$ and $x^+ = \max_{m \in \mathcal{M}} a_{m1}$.

Example 6.2 *Consider four existing facilities with the following coordinates and with equal weights:*

$$Ex_1 = (1,1), Ex_2 = (2,0), Ex_3 = (4,2), \text{ and } Ex_4 = (0,3).$$

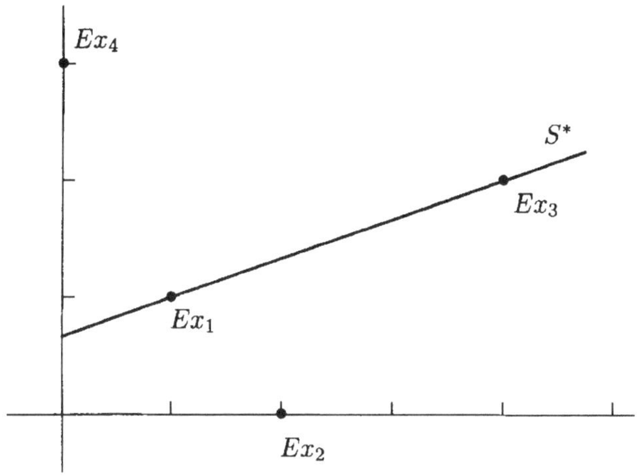

Figure 6.5. An optimal line segment S^* for the median problem.

Furthermore, choose $l_0 = 5$. We want to determine a median line segment with length 5. In the dual space we have to minimize

$$\min f(s,b),$$

$$s.t.\ |s| \leq \sqrt{\left(\frac{l_0}{span}\right)^2 - 1} = \frac{3}{4}.$$

The two vertical lines defining the two restricting sets are depicted in Figure 6.4. The optimal solution in this example is

$$X^* = (\frac{1}{3}, \frac{2}{3}),$$

which is feasible such that we get $S^* = S_{s^*,b^*,x^-}$ with $s^* = \frac{1}{3}$, $b^* = \frac{2}{3}$, and $x^- = 0$ as one optimal line segment (shown in Figure 6.5). (If all optimal solutions were forbidden then the candidate set for the median problem would be the set of intersection points of the two vertical lines and the lines L_m, $m \in \mathcal{M}$.)

6.5 Other Types of Restrictions

In Sections 6.1, 6.2, and 6.3 we investigated restrictions which are caused by a forbidden connected region R. Now we extend the theory developed for such forbidden sets to some other types of restrictions. For these extensions we can apply the same methods as in the previous sections.

Problems with a connected enforced region: Consider a connected polygon F called **enforced region** which must be met by the line facility (e.g., a new railway line must pass through some specified region close to an industrial area or nearby a town).

Note that, opposite to point location problems, such an enforced region is usually not equivalent to the restricting set $R := \mathbb{R}^2 \setminus F$. We consider again the dual version of this problem. Let $F_l = \{\text{lines } l : l \cap F \neq \emptyset\}$ denote the set of feasible lines. In the dual space we obtain a restricting set $R^D = \mathbb{R}^2 \setminus T(F_l)$ which consists of two disjoint, convex connected components R_1^D and R_2^D. If all optimal solutions $\mathcal{X}^*$ in the dual space are forbidden then $\mathcal{X}^* \subseteq R_1^D$ or $\mathcal{X}^* \subseteq R_2^D$, because $\mathcal{X}^*$ is a convex set. Thus we can replace R^D either by R_1^D or by R_2^D. Since both sets are convex we obtain an even smaller candidate set in this case, because the facets (corresponding to inner vertices) need not be evaluated. This can be extended again to all distances d derived from norms, such that we conclude the following result.

For line location problems with one enforced set F there exists an optimal line

for the median problem: which passes through one of the existing facilities and a vertex of F, or which passes through two of the existing facilities.

for the center problem: which is at maximum distance from two of the existing facilities and passes through a vertex of F, or which is at maximum distance from three of the existing facilities.

The restricting set is not connected: Suppose we are given two or more connected polygons $R_1, R_2, \ldots, R_K$ that must not be intersected by any line. To solve such problems, we first investigate the case of the vertical distance d_{ver}. We transform

$$R = \bigcup_{k=1,\ldots,K} R_k$$

to the dual space and get

$$\begin{aligned} R^T = T(R_l) &= \{X : T^{-1}(X) \cap R \neq \emptyset\} \\ &= \bigcup_{k=1,\ldots,K} \{X : T^{-1}(X) \cap R_k \neq \emptyset\} \\ &= \bigcup_{k=1,\ldots,K} (R_k)^T \end{aligned}$$

as a restricting polygonal set which may contain holes. Although such a set is not a bumpy set in the sense of Definition 15, the theory developed in Section 1.5 can also be applied to the case of polygonal sets with holes as restricting sets. In this case, all vertices of the holes have to be added to the candidate set. In our case, we have to add the intersection points of two boundaries $\partial(R_k)^T \cap \partial(R_j)^T$, for all $k, j \in \{1, 2, \ldots, K\}$ which correspond to common tangents of the two sets R_k and R_m. Summarizing, we get the following result for the vertical distance d_{ver}.

There exists an optimal line for the restricted problem which contains a facet of $conv(R_k)$ for a $k \in \{1, 2, \ldots, K\}$, or which is a common tangent of two sets R_k, R_j, $k, j \in \{1, 2, \ldots, K\}$, or

for the median problem: which passes through one of the existing facilities and is tangent to one of the sets $R_k, k \in \{1, 2, \ldots, K\}$, or which passes through two of the existing facilities.

for the center problem: which is at maximum distance from two of the existing facilities and is tangent to one of the sets $R_k, k \in \{1, 2, \ldots, K\}$, or which is at maximum distance from three of the existing facilities.

The same statement holds also for t-distances and in the case of block norms. But since Lemma 6.4 is not true if R^T is a polygonal set with holes, a generalization to all distances derived from norms like in Section 6.3 is not possible. In particular, there may exist optimal lines which do not satisfy one of the above properties.

Problems with more than one enforced region: Now we assume that K polygons $F_1, F_2, \ldots, F_K$ must be met by the new facility. Transforming the set of feasible lines

$$F_l = \{\text{lines } l : l \cap F_k \neq \emptyset \text{ for all } k = 1, 2, \ldots, K\}$$

to the dual space, we can no longer assume that all the connected components of

$$R = \mathbb{R}^2 \setminus T(F_l)$$

are convex as they were in the above case with $K = 1$. $T(F_l)$ may even contain holes. Therefore the facets of all enforced sets F_k and the common tangents of two different sets F_k, F_j, $k, j \in \{1, 2, \ldots, K\}$, have to be included in the candidate set in this case. For t-distances and block norm distances we finally conclude the following statement for line location problems with enforced polygonal regions $F_1, F_2, \ldots, F_K$.

There exists an optimal line which contains a facet of one of the sets F_k, $k \in \{1, 2, \ldots, K\}$, or which is a common tangent of two sets F_k, F_j, $k, j \in \{1, 2, \ldots, K\}$, or

for the median problem: which passes through one of the existing facilities and is tangent to one of the sets F_k, $k \in \{1, 2, \ldots, K\}$, or which passes through two of the existing facilities.

for the center problem: which is at maximum distance from two of the existing facilities and is tangent to one of the sets F_k, $k \in \{1, 2, \ldots, K\}$, or which is at maximum distance from three of the existing facilities.

Line parameter restrictions: When solving the line segment problem in Section 6.4 we considered two forbidden regions R_1^D and R_2^D in the *dual* space or, equivalently, one enforced region $F^D = \mathbb{R}^2 \setminus (R_1^D \cup R_2^D)$ in the *dual* space. The difference to the restricting sets and enforced sets discussed so far is that F^D and $R^D = R_1^T \cup R_2^T$ have no representations of forbidden regions or enforced regions in the primal space, i.e., there does not exist a set R such that $R = R^T$. The sets R^D and F^D are simply restrictions on the set of parameters s, b for the line. These problems can be classified as $1l/\mathbb{R}^2/R^D/d/\cdot$ and $1l/\mathbb{R}^2/F^D/d/\cdot$, respectively. Note that in the dual space a problem with enforced region F^D is equivalent to the corresponding problem with forbidden set $R^D = \mathbb{R}^2 \setminus F^D$ and vice versa. Many restrictions of this type can be imagined.

To solve line location problems with parameter restrictions for the vertical distance, two different approaches are possible. If a convex enforced region F^D (like in Section 6.4) is given, we can solve problems of this type by linear programming. The second approach is to apply Theorem 6.1 or Theorem 1.8 on page 30.

Whereas the first approach assumes that F^D is a convex polygon, the latter approach can be applied for all sets R^D whose connected components are bumpy sets.

Both approaches can be directly generalized to all t-distances and hence also to all distances d derived from block norms. If structural results can be proven for a problem of this type, the approaches may also be extended to all distances derived from arbitrary norms applying the methodology of Section 6.3.

7

LOCATING HYPERPLANES IN NORMED SPACES

In this chapter we want to extend the results of Chapters 2, 3, and 4 to normed n-dimensional spaces. Since the methodology in $\mathbb{R}^n$ is almost the same as in the plane, this extension does not need to be discussed in all details. The results of this chapter are the basis for some joint work of Martini together with the author, see [MS97, MS98b, MS98a]. The location problems that we discuss in this chapter will now be stated. Given a weighted set of M existing facilities $\mathcal{E}x = \{Ex_1, Ex_2, \ldots, Ex_M\} \subset \mathbb{R}^n$ with

$$Ex_m = (a_{m1}, a_{m2}, \ldots, a_{mn}) \in \mathbb{R}^n \text{ and } w_m > 0,$$

we want to find a hyperplane H which minimizes either the median function f or the center function g, i.e.,

for the median problem: $(1H/\mathbb{R}^n/\cdot/\gamma/\sum)$ we want to minimize

$$f(H) = \sum_{m \in \mathcal{M}} w_m d(Ex_m, H), \text{ and}$$

for the center problem: $(1H/\mathbb{R}^n/\cdot/\gamma/\max)$ our objective function is given by

$$g(H) = \max_{m \in \mathcal{M}} w_m d(Ex_m, H).$$

As distance measure we will investigate all distances d which are derived from norms

$$\gamma : \mathbb{R}^n \to \mathbb{R}.$$

The properties which will be discussed in this chapter are summarized in Table 7.1. As mentioned in Section 1.3 we use the sharper property Cen1' instead of Cen1 in this context.

Table 7.1. Some possible properties for the location of hyperplanes

Med1	There exists a median hyperplane passing through n affinely independent existing facilities.
Med2	Every median hyperplane is pseudo-halving.
Cen1	There exists a center hyperplane at maximum (weighted) distance from $n+1$ of the existing facilities.
Cen1'	There exists a center hyperplane at maximum (weighted) distance from $n+1$ affinely independent existing facilities.
Cen2	If the weights are all equal, there exists a center hyperplane parallel to a facet of the convex hull of the existing facilities.

In the following we assume that $\mathcal{E}x$ contains $n+1$ affinely independent points, since all other cases are trivial. Namely, in these cases all median and all center hyperplanes H^* contain the whole set $\mathcal{E}x$ of existing facilities and satisfy $f(H^*) = g(H^*) = 0$.

7.1 Results for the Horizontal Distance in $\mathbb{R}^n$

7.1.1 The Problem in the Primal Space

In this section we discuss $1H/\mathbb{R}^n/\cdot/d_{hor}/\cdot$, i.e., the problem of finding a hyperplane H minimizing the sum, or the maximum, of weighted horizontal distances between the given points and H.

The horizontal distance between a point $x = (x_1, x_2, \ldots, x_n) \in \mathbb{R}^n$ and a hyperplane $H_{s_1,\ldots,s_n,b}$ (with normal vector $\mathfrak{n} = (s_1, s_2, \ldots, s_n)$) can be calculated by

$$d_{hor}(x, H_{s_1,\ldots,s_n,b}) = \begin{cases} \left|\frac{b+s_1x_1+\ldots+s_nx_n}{s_1}\right| & \text{if } s_1 \neq 0 \\ 0 & \text{if } s_1 = 0 \text{ and } s_1x_1 + \ldots + s_nx_n + b = 0 \\ \infty & \text{if } s_1 = 0 \text{ and } s_1x_1 + \ldots + s_nx_n + b \neq 0 \end{cases}$$

Note that in order to find a hyperplane H minimizing $f(H)$ we can set $s_1 = 1$, as the subsequent case analysis shows.

Case 1: If $s_1 > 0$, then $H_{s_1,\ldots,s_n,b} = H_{1,\frac{s_2}{s_1},\ldots,\frac{s_n}{s_1},\frac{b}{s_1}}$.

Case 2: If $s_1 = 0$ then $s_1 a_{k1} + \ldots + s_n a_{kn} + b \neq 0$ for at least one $k \in \mathcal{M}$, otherwise $\mathcal{E}x$ does not contain $n+1$ affinely independent points. In this case $d(Ex_k, H) = \infty$, enforcing that $f(H) = \infty$ such that H cannot be optimal.

Summarizing these statements, the following objective function suffices to describe $1H/\mathbb{R}^n/\cdot/d_{hor}/\sum$.

$$\begin{aligned} f(H_{1,s_2\ldots,s_n,b}) &= \sum_{m\in\mathcal{M}} w_m d_{hor}(Ex_m, H_{1,s_2,\ldots,s_n,b}) \\ &= \sum_{m\in\mathcal{M}} w_m |b + a_{m1} + s_2 a_{m2} + \ldots + s_n a_{mn}| \\ &= \sum_{m\in\mathcal{M}} w_m |b + \langle \mathfrak{n}, Ex_m\rangle| \text{ with } \mathfrak{n} = (1, s_2, \ldots, s_n), \end{aligned}$$

whereas for the center problem $1H/\mathbb{R}^n/\cdot/d_{hor}/\max$ we get

$$\begin{aligned} g(H_{1,s_2\ldots,s_n,b}) &= \max_{m\in\mathcal{M}} w_m |b + a_{m1} + s_2 a_{m2} + \ldots + s_n a_{mn}| \\ &= \max_{m\in\mathcal{M}} w_m |b + \langle \mathfrak{n}, Ex_m\rangle|. \end{aligned}$$

We first discuss the median problem. Before proving Med1 and Med2 for the median problem with horizontal distances, it should be mentioned that [Zem84] solved this problem by formulating it as a linear program and applying the linear time algorithms for linear programming from Megiddo [Meg84] resulting in a linear time algorithm for any fixed dimension.

Lemma 7.1 *There exists a hyperplane minimizing the sum of weighted horizontal distances*

$$f(H) = \sum_{m\in\mathcal{M}} w_m d_{hor}(Ex_m, H)$$

and passing through n affinely independent points $Ex_m \in \mathcal{E}x$.

Proof: Let us assume that $H^* = H_{1,s_2^*,\ldots,s_n^*,b^*}$ is an optimal hyperplane which does not pass through n affinely independent points $Ex_m \in \mathcal{E}x$. Define

$$\mathcal{M}_0 := \{m \in \mathcal{M} : |b^* + a_{m1} + s_2^* a_{m2} + \ldots + s_n^* a_{mn}| = 0\}.$$

By continuity of $d(Ex_m, H^*) = |b^* + a_{m1} + s_2^* a_{m2} + \ldots + s_n^* a_{mn}|$ there exists a neighbourhood $U = U(s_2^*, \ldots, s_n^*, b^*)$ with respect to $\mathbb{R}^n$ around that solution such that for all $(s_2, \ldots, s_n, b) \in U$

$$|b + a_{m1} + s_2 a_{m2} + \ldots + s_n a_{mn}| \neq 0 \text{ for all } m \notin \mathcal{M}_0.$$

Now we consider

$$\mathcal{L} := \{(s_2, \ldots, s_n, b) : |b + a_{m1} + s_2 a_{m2} + \ldots + s_n a_{mn}| = 0 \text{ for all } m \in \mathcal{M}_0\}.$$

We know that $\mathcal{L} \neq \emptyset$, since $(s_2^*, \ldots, s_n^*, b^*) \in \mathcal{L}$. Since $\{Ex_m : m \in \mathcal{M}_0\}$ is not affinely independent, the solution of the linear system

$$|b + a_{m1} + s_2 a_{m2} + \ldots + s_n a_{mn}| = 0 \text{ for all } m \in \mathcal{M}_0$$

is not unique. Thus there exists another hyperplane $H' = H_{1,s'_2,\ldots,s'_n,b'} \neq H^*$ such that $Ex_m \in H'$ for all $m \in \mathcal{M}_0$. In other words, $(s'_2, \ldots, s'_n, b') \in \mathcal{L}$. Thus, $\mathcal{L}$ is the solution space of a linear system containing at least two different points, and therefore $\dim(\mathcal{L}) \geq 1$. Since the objective function

$$f(H_{1,s_2,\ldots,s_n,b}) = \sum_{m \notin \mathcal{M}_0} w_m |b + a_{m1} + s_2 a_{m2} + \ldots + s_n a_{mn}|$$

is linear on $U \cap \mathcal{L}$, we can find a solution which is at least as good as $H_{1,s_2^*,\ldots,s_n^*,b^*}$ and passes through n affinely independent points $Ex_m \in \mathcal{E}x$.

QED

Lemma 7.2 *Every hyperplane H^* minimizing*

$$f(H) = \sum_{m \in \mathcal{M}} w_m d_{hor}(Ex_m, H)$$

is pseudo-halving.

Proof: Consider any optimal hyperplane $H^* = H_{\mathfrak{n}^*,b^*}$. For $\mathfrak{n}^* = (1, s_2^*, \ldots, s_n^*)$ we define the one-dimensional function

$$\bar{f}(b) = \sum_{m \in \mathcal{M}} w_m |b + \langle \mathfrak{n}^*, Ex_m \rangle|.$$

Since H^* is optimal we conclude that b^* minimizes $\bar{f}$, because

$$f(H_{\mathfrak{n}^*,b}) = \bar{f}(b) \text{ holds for all } b \in \mathbb{R}.$$

From Lemma 1.9 we therefore know that b^* is a median of the weighted set $\{(- < \mathfrak{n}, Ex_m >, w_m)\}$. Defining for b^* the two sets

$$\begin{aligned} \mathcal{M}^+ &= \{m \in \mathcal{M} : b^* + \langle \mathfrak{n}^*, Ex_m \rangle > 0\} \text{ and} \\ \mathcal{M}^- &= \{m \in \mathcal{M} : b^* + \langle \mathfrak{n}^*, Ex_m \rangle < 0\}, \end{aligned}$$

we conclude that both inequalities

$$\begin{aligned} \sum_{m \in \mathcal{M}^+} w_m &\leq \frac{1}{2} W \text{ and} \\ \sum_{m \in \mathcal{M}^-} w_m &\leq \frac{1}{2} W \end{aligned}$$

hold, which equivalently means that H_{n^*,b^*} is pseudo-halving.

QED

Now let us turn to the center problem with horizontal distances, i.e., we consider $1H/\mathbb{R}^n/\cdot/d_{hor}/\max$. This problem is reducible to a linear program with $n+1$ variables, and so one gets an $O(M)$ time algorithm for any fixed dimension using the linear programming technique of [Meg84]. Using the objective function derived on page 141, the linear program reads as follows.

$$\begin{array}{lrcl} & & \min z & \\ s.t. & b + a_{m1} + s_2 a_{m2} + \ldots + s_n a_{mn} & \leq & z \text{ for all } m \in \mathcal{M} \\ & -b - a_{m1} - s_2 a_{m2} + \ldots - s_n a_{mn} & \leq & z \text{ for all } m \in \mathcal{M}. \end{array}$$

Now we discuss Cen1, Cen1', and Cen2 for center hyperplanes.

Lemma 7.3 *There exists a hyperplane minimizing the maximum of weighted horizontal distances*

$$g(H) = \max_{m \in \mathcal{M}} w_m d_{hor}(Ex_m, H)$$

which is at maximum distance from n+1 affinely independent existing facilities.

Proof: Defining M linear functions

$$g_m(s_2, s_3, \ldots, s_n) = d_{hor}(Ex_m, H_{1,s_2,\ldots,s_n,b})$$

for all $m \in \mathcal{M}$, we can formulate the problem of finding a center hyperplane in the usual form of a convex minmax optimization problem in n variables, i.e.,

$$\min \max_{m \in \mathcal{M}} g_m(s_2, \ldots, s_n, b).$$

Thus we can apply a theorem of [Drc82] (based on Helly's Theorem, see [Hel23]) which states that for convex minmax problems there always exists a solution which is only determined by $n+1$ different functions g_m, i.e., there exists an optimal hyperplane H^* and a subset $\mathcal{M}^0 \subseteq \mathcal{M}$ with $|\mathcal{M}^0| = n+1$ such that H^* also minimizes

$$\max_{m \in \mathcal{M}^0} d_{hor}(Ex_m, H^*).$$

Consequently H^* is at maximum distance from $n+1$ of the existing facilities, namely from all points Ex_m with $m \in \mathcal{M}^0$.
It remains to show that the points Ex_m with $m \in \mathcal{M}^0$ are affinely independent. Suppose that they are not affinely independent, i.e., there exists a hyperplane H containing $\{Ex_m : m \in \mathcal{M}^0\}$. Using that $g(H^*) > 0$ (since we assumed that $\mathcal{E}x$ contains at least n+1 affinely independent points) we calculate

$$\max_{m \in \mathcal{M}^0} d_{hor}(Ex_m, H) = 0 < g(H^*) = \max_{m \in \mathcal{M}^0} d_{hor}(Ex_m, H^*).$$

This is a contradiction to the optimality of H^* with respect to $\mathcal{M}^0$. QED

From the proof of Lemma 7.3 we can directly conclude the following (sharper) Corollary.

Corollary 7.1 *There exists an optimal hyperplane H^* for $1H/\mathbb{R}^n/\cdot/d_{hor}/\max$ and a subset $\mathcal{M}^0 \subseteq \mathcal{M}$ of cardinality $n+1$ (satisfying that $\{Ex_m : m \in \mathcal{M}^0\}$ is an affinely independent set) such that H^* is also optimal for the smaller problem with only $n+1$ existing facilities, namely for*

$$\min \max_{m \in \mathcal{M}^0} w_m d_{hor}(Ex_m, H).$$

While we will be able to extend Lemma 7.3 to all distances derived from norms in Section 7.2.3, this is not possible for Corollary 7.1. The following example taken from [Bar97] shows that this is not possible even in the planar case.

Example 7.1 *Consider $1l/\mathbb{R}^2/\cdot/l_2/\max$ with four existing facilities with equal weights, located as the vertices of a rectangle, e.g.,*

$$Ex_1 = (0,0), Ex_2 = (2,0), Ex_3 = (0,1), \text{ and } Ex_4 = (2,1).$$

The optimal line minimizing

$$g(l) = \max_{m=1}^{4} l_2(Ex_m, l)$$

is $l = \{(x_1, x_2) : x_2 = \frac{1}{2}\}$, and l has maximum distance to all four existing facilities. On the other hand, no subset of cardinality 3 from $\mathcal{E}x$ determines the line l, e.g., the line minimizing

$$g(l) = \sum_{m=1,2,3} l_2(Ex_m, l)$$

is the line $l_{-1,-2,1}$, i.e., the line with slope $-\frac{1}{2}$ and intercept $\frac{1}{2}$.

At the end of this section we show that Cen2 does in general not hold for horizontal distances in $\mathbb{R}^n$ with $n \geq 3$.

Example 7.2 *Consider the following set $\mathcal{E}x$ of four existing facilities with equal weights:*

$$\begin{aligned} Ex_1 &= (0,0,1), \\ Ex_2 &= (0,0,-1), \\ Ex_3 &= (2,1,0), \text{ and} \\ Ex_4 &= (2,-1,0). \end{aligned}$$

Let H_{klj} be the hyperplane passing through the points Ex_k, Ex_l, and Ex_j. Then

$$d_{hor}(H_{123}, Ex_4) = d_{hor}(H_{124}, Ex_3) = d_{hor}(H_{134}, Ex_2) = d_{hor}(H_{234}, Ex_1) = 4,$$

such that by translating these hyperplanes the best objective value is equal to $\frac{1}{2}4 = 2$, i.e., all hyperplanes H parallel to a facet of the convex hull of $\mathcal{E}x$ have objective value $g(H) \geq 2$. On the other hand, the hyperplane

$$H_{1,0,0,-1} = \{(x_1, x_2, x_3) : x_1 = 1\}$$

satisfies $g(H) = 1$ and therefore is better than all hyperplanes parallel to facets of $conv(\mathcal{E}x)$.

7.1.2 A Dual Interpretation in n dimensions

In $\mathbb{R}^n$ there obviously exist many possible transformations from points to hyperplanes and vice versa. Since we discussed the horizontal distance d_{hor} in the last subsection, we are now looking for a transformation T^n which "preserves horizontal distances". But even with this constraint, several transformations are possible. As shown in the previous section, it is sufficient to investigate only hyperplanes which are not parallel to the x_1-axis, i.e., hyperplanes $H_{1,s_2,\ldots,s_n,b}$. This means that the transformation T^n only needs to be defined for hyperplanes H with $s_1 = 1$. It is convenient to define T^n as follows.

Definition 26 *We define the following* **transformation T^n.**
Let $x = (x_1, x_2, \ldots, x_n)$ be a point and let $H_{1,s_2,\ldots,s_n,b}$ be a non-horizontal hyperplane.

$$\begin{aligned} T^n(x) &:= H_{1,x_2,\ldots,x_n,x_1} \\ T^n(H_{1,s_2,\ldots,s_n,b}) &:= (b, s_2, s_3, \ldots, s_n) \end{aligned}$$

Note that for this transformation we have $(T^n)^{-1} \equiv T^n$.
In the plane this means that T^2 maps a point $x = (x_1, x_2)$ to the non-horizontal line l_{1,x_2,x_1} with slope $-\frac{1}{x_2}$ and intercept x_1. The transformation T given in Definition 16 on page 36, on the other hand, maps the point $x = (x_1, x_2)$ to the non-vertical line $l_{-x_1,-1,x_2}$ with slope $-x_1$ and intercept x_2. The latter transformation, T, is a special case of the following transformation $\tilde{T}^n$ defined for points $x = (x_1, x_2, \ldots, x_n)$ and non-vertical hyperplanes $H_{s_1,\ldots,s_{n-1},-1,b}$.

$$\begin{aligned} \tilde{T}^n(x) &:= H_{-x_1,\ldots,-x_{n-1},-1,x_n} \\ \tilde{T}^n(H_{s_1,\ldots,s_{n-1},-1,b}) &:= (s_1, s_2, \ldots, s_{n-1}, b). \end{aligned}$$

We remark that $\tilde{T}^n$ preserves the vertical distance in $\mathbb{R}^n$, and that $\tilde{T}^2 \equiv T$.

We will indicate points in the dual space by capital letters $X, Y, \ldots$. As we required, the transformation T^n preserves the horizontal distance.

Lemma 7.4 *Let $x \in \mathbb{R}^n$ be a point and H be a hyperplane. Then*

$$d_{hor}(x, H) = d_{hor}(T^n(H), T^n(x)).$$

In particular, $x \in H \Leftrightarrow T^n(H) \in T^n(x)$.

Proof:

$$\begin{aligned} d_{hor}(x, H) &= |x_1 + s_2 x_2 \ldots + s_n x_n + b| \\ &= d_{hor}((b, s_2, \ldots, s_n), H_{1, x_2, \ldots, x_n, x_1}) \\ &= d_{hor}(T^n(H), T^n(x)). \end{aligned}$$

QED

Thus we can transform the problem of locating a hyperplane with respect to a given point set $\mathcal{E}x \subset \mathbb{R}^n$ into the problem of locating a point in $\mathbb{R}^n$ with respect to a given set of hyperplanes $\{T^n(Ex_m) : m \in \mathcal{M}\}$. This observation is formulated in the next theorem.

Theorem 7.1 *The problem of locating a hyperplane minimizing the weighted sum (maximum) of horizontal distances to a given set of points $\mathcal{E}x$ is equivalent to the problem of locating a point in $\mathbb{R}^n$ minimizing the weighted sum (maximum) of horizontal distances to a given set of hyperplanes $T^n(\mathcal{E}x)$, i.e.,*

- $1H/\mathbb{R}^n/\cdot/d_{hor}/\sum$ *and* $1/\mathbb{R}^n/\mathcal{E}x = \{T^n(Ex_1), \ldots, T^n(Ex_M)\}/d_{hor}/\sum$ *are equivalent, and,*
- $1H/\mathbb{R}^n/\cdot/d_{hor}/\max$ *and* $1/\mathbb{R}^n/\mathcal{E}x = \{T^n(Ex_1), \ldots, T^n(Ex_M)\}/d_{hor}/\max$ *are equivalent.*

Thus for X being a point in the dual space, we can rewrite f and g as follows.

$$f(X) = \sum_{m \in \mathcal{M}} w_m d_{hor}(X, T^n(Ex_m)) \text{ and}$$

$$g(X) = \max_{m \in \mathcal{M}} w_m d_{hor}(X, T^n(Ex_m)).$$

X is a minimizer of one of these objective functions if and only if $(T^n)^{-1}(X)$ is an optimal hyperplane for the corresponding hyperplane location problem. The set of optimal solutions in the dual space is again denoted by $\mathcal{X}^*$. Now we can extend Lemma 4.2 first to median hyperplanes and then to the case of the center objective function in n dimensions. These results will be needed in Section 7.4.

Lemma 7.5 *If* $1H/\mathbb{R}^2/\cdot/d_{ver}/\sum$ *has two distinct optimal hyperplanes then it also has two non-parallel optimal hyperplanes.*

Proof: We use the dual interpretation of the problem. Let $H_1 = H_{\mathfrak{n},b^1}$ and $H_2 = H_{\mathfrak{n},b^2}$ be two optimal hyperplanes with the same normal vector $\mathfrak{n} = (1, s_2, \ldots, s_n)$. According to Theorem 1.3 we know that the set of optimal solutions

- either consists of a vertex of a cell in the dual space; in this case it is unique and no two distinct hyperplanes H_1 and H_2 can exist,

- or it is an n-dimensional cell, then there exist two optimal hyperplanes H^1, H^2 with distinct normal vectors as required,

- or it is an r-dimensional face, $0 < r < n$, yielding that there exists at least one construction hyperplane $T^n(Ex_m)$ containing both points $T^n(H_1) = (b^1, s_2, \ldots, s_n)$ and $T^n(H_2) = (b^2, s_2, \ldots, s_n)$. Thus the following two equations must be satisfied, namely

$$\begin{aligned} b^1 + a_{m2}s_2 + \ldots + a_{mn}s_n + a_{m1} &= 0 \text{ and} \\ b^2 + a_{m2}s_2 + \ldots + a_{mn}s_n + a_{m1} &= 0, \end{aligned}$$

  forcing that $b^1 = b^2$ which is a contradiction to $H_1 \neq H_2$. Hence there exist two optimal hyperplanes H^1 and H^2 with normal vectors having different directions.

QED

We cannot use the same proof for the center problem, since there may be vertical construction hyperplanes; even in the plane vertical line segments may occur as cell facets, see Figures 2.5 and 2.6. Therefore we have to establish a different (and as it turns out much simpler) argument for the following lemma.

Lemma 7.6 *If* $1H/\mathbb{R}^2/\cdot/d_{ver}/\max$ *has two distinct optimal hyperplanes, then it also has two non-parallel optimal hyperplanes.*

Proof: Choose an arbitrary optimal hyperplane $H^* = H_{n,b}$ and choose two existing facilities Ex_1 and Ex_2 lying in the two different halfspaces separated by H^*, i.e.,

$$Ex_1 \in \mathcal{B}^+_{H^*}, Ex_2 \in \mathcal{B}^-_{H^*},$$

such that they are both at maximum distance from H^*, i.e.,

$$\begin{aligned} g(H^*) &= w_1 d_{hor}(Ex_1, H^*) \\ &= w_2 d_{hor}(Ex_2, H^*). \end{aligned}$$

Two facilities satisfying this condition exist, because if all existing facilities in the same halfspace would have weighted distance less than $g(H^*)$, then moving the hyperplane parallel towards the other halfspace would strictly improve the objective value, contradicting the optimality of H^*. Now consider any parallel hyperplane $H = H_{n,b+\epsilon}$ with $\epsilon \in \mathbb{R}$. We observe that

$$g(H) \in \{g(H^*) + w_1\epsilon, g(H^*) + w_2\epsilon\},$$

since moving H^* either increases the distance to Ex_1, or it increases the distance to Ex_2. We conclude that there cannot exist two optimal hyperplanes with the same normal vector.

QED

Note that the result for the center problem is much stronger as that given in Lemma 7.5, since in the median case there may exist different parallel hyperplanes which are both optimal. The reason for this difference between the median and the center problem can even be explained in the planar case. In $\mathbb{R}^2$, the mid-line in the center case cannot have full dimensional pieces, whereas the pointwise median of the set of hyperplanes is not necessarily unique. This means that the mid-line only consists of line segments, but the "median-line", i.e., the pointwise median of $L_1, \ldots, L_M$ can include complete cells.

Corollary 7.2 *Let $H^* = H_{\mathfrak{n},b}$ be a center hyperplane (for $1H/\mathbb{R}^n/\cdot/d_{hor}/\max$). Then no other hyperplane $H = H_{\mathfrak{n},b'}$ with the same normal vector $\mathfrak{n}$ is optimal.*

7.2 Results in Normed Spaces

7.2.1 The Rectangular Distance in $\mathbb{R}^n$

An extension of the previous results to the case of the rectangular distance is directly possible by using the results for the horizontal distance d_{hor}. Defining the distances in the other directions $e_2, e_3, \ldots, e_n$ as

$$d_{e_i}(x, H) = \begin{cases} \left|\frac{b+s_1x_1+\ldots+s_nx_n}{s_i}\right| & \text{if } s_i \neq 0 \\ 0 & \text{if } s_i = 0 \text{ and } s_1x_1 + \ldots + s_nx_n + b = 0 \\ \infty & \text{if } s_i = 0 \text{ and } s_1x_1 + \ldots + s_nx_n + b \neq 0, \end{cases}$$

we obtain the same results as in the horizontal case $d_{hor}(x, H) = d_{e_1}(x, H)$. The rectangular distance between a point $x_m \in \mathbb{R}^n$ and a hyperplane $H = H_{s_1,s_2,\ldots,s_n,b}$ is given by

$$\begin{aligned} l_1(x, H) &= \min_{i=1,2,\ldots,n} \frac{1}{|s_i|}|b + s_1x_1 + \ldots + s_nx_n| \\ &= \min_{i=1,2,\ldots,n} d_{e_i}(x, H). \end{aligned}$$

Thus Lemma 7.1, Lemma 7.2, and Lemma 7.3 show that Med1, Med2, and Cen1' hold also for the rectangular distance $d = l_1$.

7.2.2 The Euclidean Distance in $\mathbb{R}^n$

For $n \geq 3$, it was already mentioned in [NM80] that there *exists* a median hyperplane spanned by n affinely independent given points. Confirming this incidence criterion and using it together with basic techniques from computational geometry, [HII$^+$93] obtained an $O(M^n)$ worst-case time algorithm that finds *one* optimal hyperplane. Independently, [KM90] obtained the same time and space complexity, but on the much stronger geometrical basis that *each* median hyperplane has to pass through n affinely independent given points, see also [KM93]. (Thus, *every* median hyperplane can be obtained in $O(M^n)$ time.) Unfortunately, until now no higher-dimensional analogue to the computational evaluation of the line rotating procedure of [Lov71], see Section 3.1 on page 49, is known, although existing upper bounds on the number of halving hyperplanes (cf. [DE94] for $n = 3$ and [ZV92] for $n \geq 4$) give some hope for improvements in the unweighted case. (In the weighted situation, it even remains to be answered if cM^n is the worst case number of halving hyperplanes.)

For the center problem in the Euclidean plane, [MN83] confirmed Cen1. Their proof can easily be extended to dimensions $n \geq 3$. Using a reformulation of the problem with linear constraints, [HII+93] succeeded to develop an $O(M^{\lfloor \frac{n}{2}+1 \rfloor})$ worst case time algorithm for all dimensions $n \geq 3$.
Summarizing the above discussion we note that Med1 and Cen1 hold for hyperplane location problems with Euclidean distance.

7.2.3 *Arbitrary Norms in $\mathbb{R}^n$*

In this section we extend the results of Section 7.1 to all distances derived from norms. Since the method is the same as in the planar case (see Section 3.2), some proofs can be omitted. The main difference is that instead of the slope $s \in \mathbb{R}$ of the line we now use the normal vector $\mathfrak{n}$ of the hyperplane. This leads to some modifications in the proofs, but the results of this section are more or less similar to those in the plane. In particular, we will prove that Med1, Med2, and Cen1' hold for all distances derived from norms in $\mathbb{R}^n$ by transferring the corresponding results holding for d_{hor} to the case of arbitrary norms γ.

As in the planar case, we note that in order to determine the distance between a point x and a hyperplane H we can dilate the unit ball with respect to x until it is supported by the hyperplane. This immediately yields the following lemma, which can be proved analogously to Lemma 3.7.

Lemma 7.7 *For any norm γ with unit ball B and the derived distance d, any hyperplane H, and any point $x \in \mathbb{R}^n$ the following equality holds:*

$$d(x, H) = \min\{|\lambda| : (x + \lambda B) \cap H \neq \emptyset\}.$$

We need again the definition of the t-distance, which we define in this case as a distance between points and hyperplanes.

Definition 27 *Let $t \in \mathbb{R}^n$ be a given direction. For $x \in \mathbb{R}^n$ and any hyperplane $H \subset \mathbb{R}^n$ we define the* **t-distance** *as*

$$d_t(x, H) := \min\{|\lambda| : x + \lambda t \in H\},$$

where $\min \emptyset := \infty$.

As in the case of $\mathbb{R}^2$, it can be shown that the t-distance between a point and a hyperplane can be derived from the following distance between two points $x, y \in \mathbb{R}^n$:

$$d_t(x, y) := \gamma_t(y - x),$$

where

$$\gamma_t(x) := \begin{cases} |\alpha| & \text{if } x = \alpha t \\ \infty & \text{otherwise.} \end{cases}$$

Thus we get $d_t(x, H) = \min_{z \in H} d_t(x, z)$ as in the case of $\mathbb{R}^2$. But note that in the higher dimensional case $d_t(x, H) < \infty$ for all $x \in \mathbb{R}^n$ if and only if t is not orthogonal to the normal vector $\mathfrak{n}$ of H, i.e., $\langle \mathfrak{n}, t \rangle \neq 0$. Now it is easy to verify that Lemma 3.9 holds in $\mathbb{R}^n$, simply by replacing $\mathbb{R}^2$ by $\mathbb{R}^n$ in its proof.

Lemma 7.8 *Let $p, q \in \mathbb{R}^n$ and D be a linear transformation with $D(p) = q$ and $det(D) \neq 0$. Then*

$$d_q(D(x), D(H)) = d_p(x, H),$$

where $D(H) := \{D(y) : y \in H\}$.

We can also extend Med1, Med2, and Cen1' to the case of distances d_t.

Theorem 7.2 *For all distances d_t the following holds:*

Med1: *There exists a median hyperplane which passes through n affinely independent points $Ex_m \in \mathcal{E}x$.*

Med2: *All median hyperplanes are pseudo-halving ones.*

Cen1': *There exists a center hyperplane which is at maximum distance from $n+1$ affinely independent existing facilities.*

Proof: To solve $1H/\mathbb{R}^n/\cdot/d_t/\cdot$ we choose

$$\begin{aligned} q &= e_1, \\ p &= t, \end{aligned}$$

and transform the problem with t-distance to the corresponding problem of the type $1H/\mathbb{R}^n/\cdot/d_{hor}/\cdot$ according to Lemma 7.8. Then we know that H is optimal for the problem $1H/\mathbb{R}^n/\cdot/d_t/\cdot$ if and only if $D(H)$ is optimal for $1H/\mathbb{R}^n/\cdot/d_{hor}/\cdot$.

Solving $1H/\mathbb{R}^n/\cdot/d_{hor}/\sum$ yields an optimal solution H^*_{hor} which passes through n affinely independent points $D(Ex_m)$, for all $m \in \mathcal{M}$ (Lemma 7.1). We also know from Lemma 7.2 that all optimal hyperplanes are pseudo-halving ones for the transformed point set $D(\mathcal{E}x) = \{D(Ex_m) : m \in \mathcal{M}\}$.

For the solution of the center problem $1H/\mathbb{R}^n/\cdot/d_{hor}/\max$ we can choose an optimal hyperplane H^*_{hor} which is at maximum distance from $n+1$ affinely independent existing facilities (see Lemma 7.3).

Determine $H_t = D^{-1}(H^*)$. Med2 then also holds for the original problem $1H/\mathbb{R}^n/\cdot/d_t/\sum$ (compare the proof of Lemma 3.9), and since affine independence is invariant with respect to the transformation D, Med1 and Cen1' also hold for the original problem, completing the proof.

QED

The result of the next lemma extends the result of Lemma 3.10 to dimensions $n \geq 3$.

Lemma 7.9 *Let γ be a norm or $\gamma = \gamma_t$ for some vector $t \in \mathbb{R}^n$ and let $d(x,y) = \gamma(y-x)$ be the corresponding distance. Let $\mathfrak{n} \in \mathbb{R}^n$ be such that t is not orthogonal to $\mathfrak{n}$. Then there exists a constant $\tilde{C} := \tilde{C}(\mathfrak{n}, d, l_2)$ such that for all $z \in \mathbb{R}^n$ and all $x \in \mathbb{R}^n$*

$$d(x, H_{\mathfrak{n},z}) = \tilde{C}\ l_2(x, H_{\mathfrak{n},z}).$$

Proof: First consider the case $x = 0$. Let $\mathfrak{n} \in \mathbb{R}^n$ be a normal vector. Choose a hyperplane H^0 with normal vector $\mathfrak{n}$ which does not contain the origin, i.e.,

$$H^0 = H_{\mathfrak{n},b_0} \text{ with } b_0 \neq 0.$$

We then know that $l_2(0, H_{\mathfrak{n},b_0}) > 0$ and $0 < d(0, H_{\mathfrak{n},b_0}) < \infty$, and therefore we can find a real number $\tilde{C} > 0$ such that

$$d(0, H_{\mathfrak{n},b_0}) = \tilde{C} l_2(0, H_{\mathfrak{n},b_0}).$$

Now choose any other hyperplane $H = H_{\mathfrak{n},b}$ with normal vector $\mathfrak{n}$.
Since $b_0 \neq 0$, there exists β such that $b = \beta b_0$. Now we investigate the new hyperplane $\beta H_{\mathfrak{n},b_0}$.

$$\begin{aligned}
\beta H_{\mathfrak{n},b_0} &= \{\beta x : x \in H_{\mathfrak{n},b_0}\} \\
&= \{\beta x : \langle \mathfrak{n}, x\rangle + b_0 = 0\} \\
&= \{x' : \langle \frac{1}{\beta}\mathfrak{n}, x'\rangle + b_0 = 0\} \\
&= \{x' : \langle \mathfrak{n}, x'\rangle + \beta b_0 = 0\} \\
&= H_{\mathfrak{n},\beta b_0} = H_{\mathfrak{n},b}.
\end{aligned}$$

Since this transfers the equation $\beta l_{s_1,s_2,b} = l_{s_1,s_2,\beta b}$, holding for lines in the plane, to hyperplanes in $\mathbb{R}^n$, we can proceed as in the proof of Lemma 3.10 and get

$$|\beta| d(0, H_{\mathfrak{n},b_0}) = d(0, H_{\mathfrak{n},b}).$$

Therefore we can conclude that

$$d(0, H_{\mathfrak{n},b}) = |\beta| d(0, H_{\mathfrak{n},b_0}) = |\beta| \tilde{C} l_2(0, H_{\mathfrak{n},b_0}) = \tilde{C} l_2(0, H_{\mathfrak{n},b}),$$

using the above equation for both d and l_2. Since d is derived from a norm and is therefore translation invariant in $\mathbb{R}^n$, we finally get $d(x, H_{\mathfrak{n},b}) = \tilde{C} l_2(x, H_{\mathfrak{n},b})$ for any point $x \in \mathbb{R}^n$.

QED

l_2 can be replaced by any other distance derived from a norm or by distances derived from γ_t with t and $\mathfrak{n}$ not orthogonal, as in the case of the constant C in the plane. Formally, the relation between C and $\tilde{C}$ (if d_1 and d_2 are distances derived from norms, or t-distances in the plane) is the following.

$$\begin{aligned}\tilde{C}((s_1,s_2),d_1,d_2) &= C(-\frac{s_1}{s_2},d_1,d_2) \text{ and}\\ C(s,d_1,d_2) &= \tilde{C}((s,-1),d_1,d_2).\end{aligned}$$

Example 7.3 *If $H_{s,-1,b} = \{(x_1,x_2) : x_2 = sx_1 + b\}$ is a non-vertical line in the plane with $s \neq 0$, we consequently obtain*

$$\tilde{C}((s,-1),d_{hor},d_{ver}) = |s|.$$

Another example is given by the following relation, holding for a hyperplane $H := H_{s_1,\ldots,s_n,b} = H_{\mathfrak{n},b}$ with $\langle \mathfrak{n}, e_1\rangle \neq 0$ (compare Section 7.1).

$$\tilde{C}(\mathfrak{n},l_1,d_{hor}) = \min_{i=1,2,\ldots,n} \frac{s_1}{s_i}.$$

As an immediate consequence of Lemma 7.9 the following independence of norm result holds for optimal hyperplanes with given slope.

Corollary 7.3 *Let d be a distance derived from a norm.*

1. *For a given $\mathfrak{n} \in \mathbb{R}^n$ the median hyperplanes H with normal vector $\mathfrak{n}$, i.e., the hyperplanes $H_{\mathfrak{n},z^*}$ minimizing $f(H_{\mathfrak{n},z})$, are independent of the distance measure d.*

2. *For a given $\mathfrak{n} \in \mathbb{R}^n$ the center hyperplanes H with normal vector $\mathfrak{n}$, i.e., the hyperplanes $H_{\mathfrak{n},z^*}$ minimizing $g(H_{\mathfrak{n},z})$, are independent of the distance measure d.*

The next two lemmas are extensions of the corresponding results (Lemmas 3.11 and 3.12) in Section 3.2. Whereas the proof of Lemma 7.10 can be adapted literally, a few modifications in the proof of Lemma 7.11 are necessary.

Lemma 7.10 *Let γ be a norm with derived distance d. For $x \in \mathbb{R}^n$ and a hyperplane H*

$$d(x,H) = \min_{t\in\mathbb{R}^n,\gamma(t)=1} d_t(x,H).$$

Lemma 7.11 *Let H be a hyperplane, and let d be a distance derived from a norm γ. Then there exists a direction $t \in \mathbb{R}^n$ such that*

$$d(x,H) = d_t(x,H) \text{ for all } x \in \mathbb{R}^n.$$

Proof: Let $x \in \mathbb{R}^n$ and $\mathfrak{n}$ be the normal vector of the hyperplane H. According to Lemma 7.10 we can find a direction $u \in \mathbb{R}^n$ such that

$$\gamma(u) = 1 \text{ and } d(x,H) = d_u(x,H) \leq d_t(x,H)$$

for all $t \in \mathbb{R}^n$. Suppose that there exist points $y \in \mathbb{R}^n$ and $v \in \mathbb{R}^n$ with

$$\gamma(v) = 1 \text{ and } d(y, H) = d_v(y, H) < d_u(y, H).$$

Note that u is not orthogonal to $\mathfrak{n}$, nor is v, since $d(x, H) \neq \infty$ and $d(y, H) \neq \infty$. By Lemma 7.9 there exists a constant $\tilde{C} := \tilde{C}(\mathfrak{n}, d_u, d_v)$ such that $d_u(x, H) = \tilde{C} d_v(y, H)$ and $d_u(x, H) = \tilde{C} d_v(y, H)$, yielding $\tilde{C} > 1$ in the first case and $\tilde{C} \leq 1$ in the second case, which is impossible.

QED

Now we have developed all tools we need to present the main result of this section.

Theorem 7.3 *For all distances d derived from norms the following criteria hold.*

Med1: *There exists a median hyperplane which passes through n affinely independent points* $Ex_m \in \mathcal{E}x$.

Med2: *All median hyperplanes are pseudo-halving ones.*

Cen1': *There exists a center hyperplane which is at maximum distance from* $n + 1$ *affinely independent existing facilities.*

Proof: Except for some algebraic derivations, the proof for the planar case only uses Lemma 3.11, Lemma 3.12, and Theorem 3.1. These results have been shown to hold also in the higher dimensional case, compare Lemma 7.10, Lemma 7.11, and Theorem 7.2. Hence the proof of Theorem 3.3 can simply be copied to prove Theorem 7.3.

QED

7.3 Algorithmic Approaches for Hyperplane Location

Lemma 7.7 states that the distance $d(x, H)$ between a point $x \in \mathbb{R}^n$ and a hyperplane H strongly depends on the shape of the unit ball B (as it has already been mentioned for the planar case). Therefore it can be difficult to evaluate $d(x, H)$. But as in the plane we ignore such numerical difficulties and assume that a norm evaluation can be done in constant time.

Under this assumption the median problem can be solved efficiently, since Theorem 7.3 yields the possibility of algorithmical approaches analogous to those in the Euclidean case. Namely, the basic incidence criteria for optimal hyperplanes coincide (see Lemma 6.3 in [HII+93] and Theorem 2 in [KM93] for higher dimensional problems with the Euclidean distance and Med1 of Theorem 7.3 for the generalization to all distances derived from norms); the only difference is that for arbitrary norms not every median hyperplane will be necessarily found. But in order to find at least one median hyperplane, i.e. to solve $1H/\mathbb{R}^n/\cdot/\gamma/\sum$ in every n-dimensional normed space, it suffices to apply the algorithms for the Euclidean norm. One approach has been described in [KM93], pp. 138-142, and another algorithm can be found in [HII+93], pp. 227-230. With the help of both procedures we get an $O(M^n)$ time algorithm for finding a median hyperplane in $\mathbb{R}^n$.

To solve the problem of finding a center hyperplane, the Euclidean approach cannot be adapted so easily. Anyway, an enumeration of all hyperplanes which are at maximum distance from $n+1$ affinely independent existing facilities (see Cen1' in Theorem 7.3) leads to a polynomial approach for all distances derived from norms.

In the special case that the distance measure $d = d_B$ is derived from a block norm (i.e., the unit ball B is an n-dimensional polytope) it is possible to find median and center hyperplanes more efficiently, using the ideas described in Section 3.4.

Recall that the unit ball B of a block norm is a compact, convex polytope with nonempty interior and extreme points

$$Ext(B) = \{b_1, b_2, \dots, b_G, -b_1, -b_2, \dots, -b_G\}, \quad b_i \in \mathbb{R}^n, i = 1, \dots, G.$$

The next lemma is a direct extension of Lemma 3.14 on page 71.

Lemma 7.12 *Let d_B be derived from a block norm γ_B with fundamental directions $Ext(B) := \{b_1, b_2, \dots, b_G\}$ and let H be a hyperplane. Then there exists a $g \in \{1, 2, \dots, G\}$ such that for all $x \in \mathbb{R}^n$*

$$d_B(x, H) = d_{b_g}(x, H).$$

Proof: In order to determine $d_B(x, H)$ we know from Lemma 7.7 that the unit ball B of the block norm can be dilated around x until it touches H. In the case of a block norm, B is a polytope. Like in the proof of Lemma 3.14 (see page 71) we note that a hyperplane touches a polytope in at least one vertex of that polytope, such that the formula

$$d(x, H) = \min_{t \in \mathbb{R}^2, \gamma(t)=1} d_t(x, H)$$

from Lemma 7.10 can be simplified to

$$d_B(x, H) = \min_{g=1,\dots,G} d_{b_g}(x, H).$$

With Lemma 7.11 we conclude that there exists a $b_g \in Ext(B)$ such that

$$d_B(x, H) = d_{b_g}(x, H) \text{ for all } x \in \mathbb{R}^n.$$

QED

Lemma 7.12 and Lemma 7.8 are the basis for the following decomposition algorithm in $\mathbb{R}^n$.

Algorithm 11 (for solving $1H/\mathbb{R}^n/\cdot/\gamma_B/\sum$ and $1H/\mathbb{R}^n/\cdot/\gamma_B/\max$)

Input: Block norm distance d_B, $x_m, w_m > 0$ for all $m \in \mathcal{M}$

Output: Hyperplane H^* with objective value z^* which

for the median problem : solves $1l/\mathbb{R}^2/\cdot/d_B/\sum$
for the center problem : solves $1l/\mathbb{R}^2/\cdot/d_B/\max$

1. $z^* := \infty$

2. For $g = 1$ to G do

1. Determine a transformation D such that $D(b_g) = e_1$ and $det(D) \neq 0$.
2. For $m \in \mathcal{M}$ do: $Dx_m = D(Ex_m)$.
3. Find a hyperplane H_g^* minimizing
   for the median problem : $f(H) = \sum_{m \in \mathcal{M}} w_m d_{hor}(Dx_m, H)$. (Use the algorithm of Zemel [Zem84].)
   for the center problem : $f(l) = \max_{m \in \mathcal{M}} w_m d_{hor}(Dx_m, H)$. (Use the algorithm of Megiddo [Meg84].)
4. If $f(H_g^*) < z^*$, then set

$$\begin{aligned} z^* &:= f(H_g^*) \text{ and} \\ H^* &:= D^{-1}(H_g^*). \end{aligned}$$

3. Output: H^* with objective value z^*

As in the planar case, this algorithm runs in $O(GM)$ time since both corresponding problems with horizontal distances ($1H/\mathbb{R}^n/\cdot/d_{hor}/\sum$ and $1H/\mathbb{R}^n/\cdot/d_{hor}/\max$) can be solved in linear time for any fixed dimension, see the algorithm in [Zem84] for the median problem and take the linear programming methods of [Meg84] for solving the center problem.

Here, once more a connection to transversal theory becomes apparent: Algorithms solving center line or center hyperplane problems can also be used to determine hyperplane transversals. In the following we will show how Algorithm 11 can be used to solve a high-dimensional stabbing problem, i.e., to verify quickly whether a family of (not necessarily disjoint) scaled translates of a symmetric polyhedron in $\mathbb{R}^n$ has a hyperplane transversal or not.

Algorithm 12 (for finding a hyperplane transversal for a set of scaled translates of a symmetric polyhedron)

Input: A polyhedron B which is symmetric with respect to the origin, defined by its extreme points $Ext(B) = \{b_1, b_2, \ldots, b_G, -b_1, b_2, \ldots, -b_G\}$, and a family $\mathcal{A} = \{A_1, A_2, \ldots, A_M\}$ of scaled translates of B, i.e.,

$$A_m = x_m + \lambda_m B, \text{ with } x_m \in \mathbb{R}^n, \lambda_m > 0 \text{ for all } m \in \mathcal{M}.$$

Output: A hyperplane transversal for $\mathcal{A}$, if it exists

1. Define for all $m \in \mathcal{M}$

$$\begin{aligned} Ex_m &:= x_m \text{ and} \\ w_m &:= \frac{1}{\lambda_m} \end{aligned}$$

as a set of existing facilities $\mathcal{E}x$ with positive weights w_m.

2. Use Algorithm 11 in the center case to get a hyperplane H^* and an objective value z^* for the set of existing facilities defined in Step 1.

3. If $z^* \leq 1$, then H^* is a hyperplane transversal for $\mathcal{A}$. If $z^* > 1$, then no hyperplane transversal for $\mathcal{A}$ exists.

Algorithm 12 runs in $O(GM)$ time. In the plane, the best known algorithm to find a transversal of translates of a convex polytope also runs in $O(GM)$ time, see [Ede85], where G is the number of different slopes of all edges of the polytope which in our case is equal to the number of fundamental directions G. This algorithm does not assume that the translates are symmetric to an interior point as we do in Algorithm 12. On the other hand, we allow in Algorithm 12 that the sets under consideration are *scaled* translates of a common set B, i.e., it is not required that they all have the same size. This is an extension to the currently published algorithms in transversal theory.

For the higher dimensional case, however, Algorithm 12 finds a stabbing hyperplane in $\mathbb{R}^n$, and its running time is still linear in M and linear in G, independently of the dimension n.

Only few algorithms for solving high-dimensional stabbing problems have been discussed in the literature so far. In [AD88] the problem of stabbing M line segments in $\mathbb{R}^n$ by a hyperplane is solved in $O(M^n)$ time, and for $n = 3$ a plane stabber for a set of M polyhedra with a total of GM vertices can be found in $O(G^2M^3)$. Theoretical results about hyperplane transversals can be found in [PW90] and [CGP+90]. In [Nie96], the problem of stabbing boxes in higher dimensions is investigated.

Compared to these approaches, Algorithm 12 seems to be the only transversal algorithm (for polyhedra in $\mathbb{R}^n$) so far with a running time independent of the dimension n of the space.

7.4 A Characterization of Smooth Norms

Since the definition of smooth norms has so far only been given in the plane, we repeat it here for more than two dimensions.

Definition 28 (e.g. [Phe89]) *Let γ be a norm with unit ball $B = \{x \in \mathbb{R}^n : \gamma(x) \leq 1\}$. γ is called a* **smooth norm**, *if for all $t \in \mathbb{R}^n$ with $\gamma(t) = 1$ there exists exactly one supporting hyperplane of B at t.*

As in the planar case we need the following two lemmas for the proof of the main result. The proof of Lemma 7.13 is identical to the proof of Lemma 4.1 in the planar case, such that it can be omitted here.

Lemma 7.13 *Let d be a distance derived from a norm γ. Let $t, H \in \mathbb{R}^n$ be such that $\gamma(t) = 1$ and such that $d(x, H) = d_t(x, H)$ for all $x \in \mathbb{R}^n$. Then there exists a translate of H which is a supporting hyperplane of the unit ball B of γ at t.*

Lemma 7.14 *Let $t \in \mathbb{R}^n \setminus \{0\}$. If $1H/\mathbb{R}^2/\cdot/d_t/\sum$ (or $1H/\mathbb{R}^n/\cdot/d_t/\max$) has two distinct optimal hyperplanes H_1 and H_2 then it also has two non-parallel optimal hyperplanes.*

Proof: This follows directly from Lemma 7.5, Lemma 7.6, and from Lemma 7.8.

QED

Now we extend Theorems 4.1 and 4.2 to the case of more than two dimensions. Although the proof of the first direction is nearly the same as in the planar case, we will repeat it in this context, since the result seems to be quite important.

Theorem 7.4 *Let $\gamma : \mathbb{R}^n \to \mathbb{R}$ be a norm in $\mathbb{R}^n$.*

1. *γ is smooth if and only if for all problem instances of $1H/\mathbb{R}^n/\cdot/\gamma/\sum$ all optimal hyperplanes are affine hulls of n affinely independent existing facilities.*

2. *γ is smooth if and only if for all problem instances of $1H/\mathbb{R}^n/\cdot/\gamma/\max$ all optimal hyperplanes are at maximum distance from $n+1$ affinely independent existing facilities.*

Proof:

$\Longrightarrow$: Let γ be a smooth norm and $d(x, y) = \gamma(y - x)$ be the derived distance. Let $\mathcal{E}x = \{Ex_1, Ex_2, \ldots, Ex_M\}$ be a set of existing facilities with weights $w_m > 0$ for $m = 1, 2, \ldots, M$. Now we consider an optimal hyperplane H^* for $1H/\mathbb{R}^n/\cdot/\gamma/\sum$. Let $\mathfrak{n}$ be the normal vector of H^*.

Suppose that H^* does not satisfy the claimed property. We choose $t^* \in \mathbb{R}^n$ with $\gamma(t^*) = 1$ according to Lemma 7.11 such that

$$d_{t^*}(Ex_m, H^*) = d(Ex_m, H^*) \text{ for all } m \in \mathcal{M}.$$

Now we consider the corresponding location problem with distance d_{t^*} instead of d $(1H/\mathbb{R}^n/\cdot/d_{t^*}/\sum)$. For this problem, we can choose a hyperplane H^0 that is also optimal for $1H/\mathbb{R}^2/\cdot/\gamma/\sum$ and that satisfies the desired property, see Theorem 7.2. Since H^* and H^0 are both optimal hyperplanes, Lemma 7.14 yields that there exists a hyperplane H' with normal vector $\mathfrak{n}'$ such that H' is also optimal for the problem instance with distances d_{t^*} and $\mathfrak{n}^* \neq \mathfrak{n}'$. Now choose t' with $\gamma(t') = 1$ such that

$$d_{t'}(Ex_m, H') = d(Ex_m, H') \text{ for all } m \in \mathcal{M},$$

again using Lemma 7.11. Then Lemma 7.10 implies

$$d_{t^*}(Ex_m, H') \geq d_{t'}(Ex_m, H') \text{ for all } m \in \mathcal{M}.$$

Additionally, since γ is a smooth norm,

$$d_{t^*}(Ex_m, H') \neq d_{t'}(Ex_m, H') \text{ for all } m \in \mathcal{M},$$

because otherwise

$$d(Ex_m, H') = d_{t'}(Ex_m, H') = d_{t^*}(Ex_m, H')$$

and there would exist two different supporting hyperplanes of B at t^* according to Lemma 7.13, namely:

- a translate of H^*, since $d(Ex_m, H^*) = d_{t^*}(Ex_m, H^*)$, and
- a translate of H', since $d(Ex_m, H') = d_{t^*}(Ex_m, H')$.

Summarizing, we get $d_{t^*}(Ex_m, H') > d_{t'}(Ex_m, H')$ for all $m \in \mathcal{M}$ and therefore

$$\begin{aligned} f(H^*) &= \sum_{m \in \mathcal{M}} w_m d(Ex_m, H^*) \\ &= \sum_{m \in \mathcal{M}} w_m d_{t^*}(Ex_m, H^*) \\ &= \sum_{m \in \mathcal{M}} w_m d_{t^*}(Ex_m, H') \\ &> \sum_{m \in \mathcal{M}} w_m d_{t'}(Ex_m, H') \\ &= \sum_{m \in \mathcal{M}} w_m d(Ex_m, H') = f(H'), \end{aligned}$$

contradicting the optimality of H^*. For the center problem we can copy the above proof; we only have to replace all $\sum$ by max.

$\Longleftarrow$: Let γ be a non-smooth norm. Let H_a and H_b be two different supporting hyperplanes of B at $t \in \mathbb{R}^n$. Define $L = H_a \cap H_b$ as the intersection of H_a and H_b. Then L is an affine subspace with $dim(L) = n-2$, since $t \in H_a \cap H_b$. Choose $n-1$ points $x_1, \ldots, x_{n-1}$, such that $L = aff(\{x_1, \ldots, x_{n-1}\})$. Furthermore, let $x_a \in H_a \setminus L$ and $x_b \in H_b \setminus L$ such that $x_b - x_a = \lambda t$ for a $\lambda \in \mathbb{R}$.

For the median problem: Define a set of $M = n+1$ existing facilities $Ex_1, \ldots, Ex_{n-1}, Ex_a, Ex_b$ as the points $x_1, \ldots, x_{n-1}, x_a, x_b$. Furthermore, choose weights $w_m = c$ for $m = 1, \ldots, n-1$ and $w_a = w_b = 1$. If c is sufficiently large, any optimal hyperplane has to contain $Ex_1, \ldots, Ex_{n-1}$ and therefore the whole subspace L. Now consider the following three hyperplanes, all passing through L:

$$\begin{aligned}
H_1 &= aff(\{Ex_1, \ldots, Ex_{n-1}, Ex_a\}), \\
H_2 &= aff(\{Ex_1, \ldots, Ex_{n-1}, Ex_b\}), \text{ and} \\
H &= aff(\{Ex_1, \ldots, Ex_{n-1}, \frac{1}{2}(Ex_a + Ex_b)\}).
\end{aligned}$$

We calculate that

$$\begin{aligned}
f(H_1) &= w_b d(Ex_b, H_1) = \lambda, \\
f(H_2) &= w_a d(Ex_a, H_2) = \lambda, \\
f(H) &= w_a d(Ex_a, H) + w_b d(Ex_b, H) \\
&= \frac{\lambda}{2} + \frac{\lambda}{2} = \lambda,
\end{aligned}$$

yielding that H, H_1, and H_2 are all optimal. But H only passes through $n-1$ of the existing facilities.

For the center problem: Let $\delta > 0$. Choose the following set of $M = n+1$ existing facilities, all with weights $w_m = 1, m = 1, \ldots, M$.

$$\begin{aligned}
Ex_m &= x_m - \delta t, m = 1, \ldots, n-1, \\
Ex_n &= x_1 + \delta t, \text{ and} \\
Ex_{n+1} &= \frac{\lambda - \delta}{\lambda} x_a + \frac{\delta}{\lambda} x_b.
\end{aligned}$$

First note that $Ex_{n+1} \notin aff(\{Ex_1, \ldots, Ex_{n-1}, Ex_n\})$ yielding that there does not exist a hyperplane containing the whole set of existing facilities. Since H_a is a supporting hyperplane of B at t, we conclude that

$$d(Ex_m, H_a) = d_t(Ex_m, H_a) = \delta \text{ for all } m \in \mathcal{M}.$$

Hence, if we choose δ sufficiently small, H_a is optimal for this problem instance of $1H/\mathbb{R}^n/\cdot/\gamma/\max$. On the other hand, if $\delta < \frac{\lambda}{2}$, the hyperplane H passing through L and through Ex_{n+1} lies between H_a and H_b and therefore is also a supporting hyperplane of B at t. Hence

$$g(H) = \delta = g(H_a)$$

and H is also optimal. But $d(Ex_{n+1}, H) = 0$ such that H is not at maximum distance from $n+1$ of the existing facilities.

QED

8 EXTENSIONS: LOCATING OTHER OBJECTS

With the exception of Section 6.4, where we discussed the location of a line segment as an example for a restricted line location problem, we dealt with the location of lines and hyperplanes so far. A lot of extensions to these affine structures are possible: In the plane one might locate other one-dimensional facilities, e.g., line segments, paths, trees, circles, etc. The location of two-dimensional facilities such as rectangles, squares, or balls seems to be an interesting subject, too. But in the literature very little is known about the location of other objects *New*, when the distance between *New* and a point x is defined as the distance between x and its closest point in *New*, i.e.,

$$d(New, x) = \inf_{y \in New} d(x, y).$$

Two problems of this type have been discussed in the literature. The location of central line segments with given length with respect to the Euclidean distance $(1S/\mathbb{R}^2/\cdot/l_2/\max)$ has been discussed by [ILY92, AEST93, Pet97], whereas [DBM96] deals with the location of rectangular paths in the plane minimizing the sum of l_1-distances to the set of existing facilities. On the other hand, if the distance $d(x, New)$ between a point x and an area *New* is defined as the average distance between x and all points in *New*, some more literature is available, most dealing with areas as existing facilities, compare e.g. [KK91, Lov72].

In this chapter we will show in two examples how the theory developed in the previous chapters may be extended to the location of other structures. As first example we introduce a bicriteria line segment location problem, and then we deal with the location of circles. Finally we formulate the decision version of a general location problem, which refers to the location of arbitrarily shaped one-dimensional facilities in the plane. We show that some well known geometric and combinatorial problems are special cases of such a formulation.

8.1 Bicriteria Line Segment Problems

Instead of fixing the length l_0 of the new line segment facility, a bicriteria approach is possible, i.e., we want to minimize not only the median or the center objective function, but also the length $le(S)$ of the new facility, that gives an estimation on the costs of building up and operating the facility S. In networks, such a bicriteria formulation has been given by [CRC87]. For a general introduction into multicriteria optimization, see e.g. [Ste89] or [Ehr97]. In the following we will formulate the bicriteria segment location problem in the plane and show some properties of its set of efficient solutions. Details of this approach will be given in a forthcoming paper of Wiecek together with the author, see [SW98].

Let $\mathcal{S} \in \mathbb{R}^2$ be the set of all closed line segments S in the plane and let a and b be the two endpoints of a segment S. The segment length l_0 of S is defined as the Euclidean distance from a to b and is represented by the function $le(S)$ such that $l_0 = le(S)$, as already used in Section 6.4.

According to Definition 6 (see page 7) the distance between a facility $Ex_m \in \mathcal{E}x$ and a (closed) segment $S \in \mathcal{S}$, denoted by $d(Ex_m, S)$, is defined as

$$d(Ex_m, S) := \min_{x \in S} d(Ex_m, x).$$

We formulate the following two bicriteria segment location problems:
The Bicriteria Segment Median Problem

$$\min \begin{bmatrix} \sum_{m \in \mathcal{M}} d(Ex_m, S) \\ le(S) \end{bmatrix}$$

$$s.t.\ S \in \mathcal{S}$$

and
the Bicriteria Segment Center Problem

$$\min \begin{bmatrix} \max_{m \in \mathcal{M}} d(Ex_m, S) \\ le(S) \end{bmatrix}$$

$$s.t.\ S \in \mathcal{S}.$$

In order to simplify notation we formulate the (general) bicriteria segment location problem:

$$\min \begin{bmatrix} h(S) \\ le(S) \end{bmatrix}$$

$$s.t.\ S \in \mathcal{S},$$

where $h \in \{f, g\}$, and f and g are defined as the median and the center objective function (as usual). Throughout this section we use the function h whenever we refer to each of the functions f and g independently, significantly shortening the presentation.

Definition 29 *A segment $S \in \mathcal{S}$ is called an* **efficient segment** *if there is no other segment $S' \in \mathcal{S}$ such that $h(S') \leq h(S)$ and $le(S') \leq le(S)$ where at least one inequality holds strict.*

Let $Eff_h(\mathcal{S})$ denote the set of all efficient segments.

A common way of solving a bicriteria problem for efficient solutions is to solve single objective problems related to the bicriteria problem. Given our general bicriteria segment location problem we formulate its scalarizations following two approaches well known in the multicriteria literature. Due to [CC61] we define the following problem P_{h_0,l_0}:

$$\min h(S) + le(S)$$

$$\begin{aligned} s.t.\quad h(S) &\leq h_0 \\ le(S) &\leq l_0 \\ S &\in \mathcal{S}, \end{aligned}$$

where h_0,$l_0 \geq 0$ are given. According to [CC61] we get the following result.

Lemma 8.1 *A segment $S \in \mathcal{S}$ is an optimal solution of P_{h_0,l_0} for some $h_0, l_0 \geq 0$ if and only if $S \in Eff_h(\mathcal{S})$.*

According to [HC83] we have for a given $l_0 \geq 0$ the l_0-constraint problem $P_{\leq l_0}$:

$$\min h(S)$$

$$\begin{aligned} s.t\quad le(S) &\leq l_0 \\ S &\in \mathcal{S}, \end{aligned}$$

For $P_{\leq l_0}$ the following lemma holds (see [HC83]).

Lemma 8.2 *If a segment $S \in \mathcal{S}$ is a unique solution of $P_{\leq l_0}$ for some $l_0 \geq 0$, then $S \in Eff_h(\mathcal{S})$.*

For a given length l_0 of a segment we formulate the problem P_{l_0} of locating a line segment S of fixed length l_0 $(1S/\mathbb{R}^2/\cdot/d/\cdot)$ with respect to any distance measure d:

$$\min h(S)$$

$$\begin{aligned} s.t \quad & le(S) = l_0 \\ & S \in \mathcal{S}, \end{aligned}$$

Lemma 8.3 *The optimal objective values of $P_{\leq l_0}$ and of P_{l_0} coincide.*

Proof: It is clear that the objective value of $P_{\leq l_0}$ is not worse than the objective value of P_{l_0}. For the other direction we note the following: Let S^1 and S^2 be two closed line segments with $S^1 \subseteq S^2$. Then $d(Ex_m, S^1) \geq d(Ex_m, S^2)$ for all $m \in \mathcal{M}$ and hence $h(S^1) \geq h(S^2)$. We conclude that the objective value of $P_{\leq l_0}$ is not better than the objective value of P_{l_0}.

QED

In the following we will show for which values of l_0 even the sets of optimizers of the two problems $P_{\leq l_0}$ and of P_{l_0} coincide.

Let $S^*_{l_0}$ denote any optimal solution of P_{l_0} and let $h^*_{l_0} = h(S^*_{l_0})$ be the objective value of P_{l_0} (or of $P_{\leq l_0}$). Note that if $l_0 = 0$ or $l_0 = \infty$, P_{l_0} becomes the point location problem or the line location problem, respectively, and h^*_0 is the optimal objective value of the former while h^*_∞ is the optimal objective value of the latter.

Let l^S denote the straight line in $\mathbb{R}^2$ containing the segment S. Considering the segment length as the primary objective function we obtain the auxiliary problem (AP):

$$\min le(S)$$

$$\begin{aligned} s.t \quad & h(S) = h(l^S) \\ & l^S \in \mathcal{L}^*, \end{aligned}$$

where $\mathcal{L}^*$ is the set of all optimal lines solving the line location problem. Let S_{crit} be an optimal segment of AP and l_{crit} be the corresponding optimal objective value $l_{crit} = le(S_{crit})$.

Finally we define the cover of a segment S with respect to the distance d or d_t:

Definition 30 *Let d be a distance derived from a norm or let $d = d_t$ for some $t \in \mathbb{R}^2$. Let S be a line segment. Then the* **cover** *of S with respect to d is given by*

$$cover_d(S) = \{x \in \mathbb{R}^2 : \text{ there exists a } y \in S \text{ with } d(x,y) = d(x, l^S)\}.$$

Since $d_t(x,S) < \infty$ if and only if there exists a $y \in S$ with $d(x,y) = d(x,l^S)$, for $d = d_t$ the relation

$$cover_{d_t}(S) = \{x \in \mathbb{R}^2 : d_t(x,S) < \infty\}$$

holds. Now some straightforward results can be formulated.

Lemma 8.4 *If* $0 \le l_1 < l_2 \le \infty$, *then* $h^*_{l_1} \ge h^*_{l_2}$.

Proof: As mentioned in the proof of Lemma 8.3, $h(S^*_{l_1}) \ge h(S)$ holds for all segments S of fixed length l_2 which contain $S^*_{l_1}$. By definition $h(S) \ge h(S^*_{l_2})$. Together we conclude that $h(S^*_{l_1}) \ge h(S^*_{l_2})$ or, equivalently, $h^*_{l_1} \ge h^*_{l_2}$.

QED

Lemma 8.5 *If* $l_0 \ge l_{crit}$, *then* $h^*_{l_0} = h^*_{\infty}$.

Proof: From Lemma 8.4 it follows that $h^*_{l_0} \le h^*_{\infty}$ if $l_0 \ge l_{crit}$. Now let S be a segment of length l_0 and l^S be the line containing it. Then

$$h(S) \ge h(l^S) \ge h(l^*) = h^*_{\infty},$$

where $l^* \in \mathcal{L}^*$ is an optimal line. In particular, for $S^*_{l_0}$ we also have

$$h^*_{l_0} = h(S^*_{l_0}) \ge h^*_{\infty}.$$

Together, this implies that $h^*_{l_0} = h^*_{\infty}$.

QED

Corollary 8.1 *For all* $S \in \mathcal{S}$:

$$le(S) > l_{crit} \implies S \notin \mathit{Eff}_h(\mathcal{S}).$$

The next result follows directly from the definition of P_{l_0}.

Lemma 8.6 *Let* d *be a distance derived from a norm. Then for all* $S \in \mathcal{S}$

$$le(S) = l_0 \text{ and } h(S) > h^*_{l_0} \implies S \notin \mathit{Eff}_h(\mathcal{S}).$$

For the following results let us first consider the case of the vertical distance d_{ver}. Recall that

$$span(\mathcal{E}x) = \max_{m \in \mathcal{M}} a_{m1} - \min_{m \in \mathcal{M}} a_{m1}$$

has been introduced in Definition 25 on page 129. If it is clear to which set of existing facilities we refer we simply write *span* instead of $span(\mathcal{E}x)$.

Theorem 8.1 *Let* $d = d_{ver}$ *be the vertical distance and let* $\epsilon > 0$. *Then* $h^*_{l_0} > h^*_{l_0+\epsilon}$ *for all* l_0 *with* $span \le l_0 < l_{crit}$.

Proof: Let $l_{s,-1,b}$ denote the straight line with slope s and intercept b. Then the function defined as

$$\bar{h}(s) := \min_{b \in \mathbb{R}} h(l_{s,-1,b})$$

is convex, because

$$\tilde{h}(s,b) := h(l_{s,-1,b})$$

is a convex function (see [MT82] or [Sch97]). Let $\bar{s}$ be a slope minimizing $\bar{h}(s)$. Furthermore, for every segment $S \in \mathcal{S}$ with $h(S) < \infty$ Lemma 6.6 on page 130 implies that $h(S) = h(l^S)$. Now observe that for $l_0 < l_{crit}$, a segment $S^*_{l_0}$ must be contained in a line whose slope s does not minimize $\bar{h}$. Increasing l_0 to $l_0 + \epsilon$ we can choose a new segment $S' \in \mathcal{S}$ contained in a line with slope s' such that

$$\mathcal{E}x \subseteq cover_{d_{ver}}(S')$$

and

$$|s' - \bar{s}| < |s - \bar{s}|.$$

Since the function $\bar{h}(s)$ is convex, and therefore strictly monotone for all slopes not minimizing it, we get

$$\bar{h}(s') < \bar{h}(s).$$

Therefore $h(S^*_{l_0+\epsilon}) \leq h(S') < h(S^*_{l_0})$ or, equivalently, $h^*_{l_0+\epsilon} < h^*_{l_0}$.

QED

Theorem 8.1 leads to the following two corollaries.

Corollary 8.2 *Let $d = d_{ver}$ be the vertical distance. Then for all $S \in \mathcal{S}$*

$$span \leq le(S) \leq l_{crit} \text{ and } h(S) = h^*_{le(S)} \Longrightarrow S \in \mathcal{E}ff_h(\mathcal{S}),$$

i.e., all optimal solutions of P_{l_0} with $span \leq l_0 \leq l_{crit}$ are efficient segments.

Corollary 8.3 *For the vertical distance $d = d_{ver}$ and $span < l_0 < l_{crit}$ the set of optimal solutions of $P_{\leq l_0}$ coincides with the set of optimal solutions of P_{l_0}.*

Definition 31 (e.g. [Phe89]) *Let γ be a norm with unit ball B. γ is called a* **strictly convex norm** *if B is a strictly convex set, i.e., ∂B does not contain any line segment.*

To extend Theorem 8.1 to all distances derived from strictly convex norms, we need the following two lemmas.

Lemma 8.7 *Let d be a distance derived from a norm γ. For all $x \in \mathbb{R}^2$ and $S \in \mathcal{S}$*

$$d(x, S) = \min_{t \in \mathbb{R}^2, \gamma(t)=1} d_t(x, S).$$

Proof: The proof is analogous to the proof of Lemma 3.11 in Section 3.2 on page 64, but using Lemma 3.7 for line segments instead of lines, i.e., the distance between a point x and a line segment S is given by

$$d(x, S) = \min\{|\lambda| : (x + \lambda B) \cap S \neq \emptyset\}.$$

QED

Lemma 8.8 *Let d be a distance derived from a strictly convex norm γ. For every $S \in \mathcal{S}$ there exists a direction $u \in \mathbb{R}^2$ with $\gamma(u) = 1$, such that*

$$d(x, S) = d_u(x, S)$$

for all $x \in cover_d(S)$. In this case, $cover_{d_u}(S) = cover_d(S)$.

Proof: According to Lemma 3.12 (see page 65) for every distance d derived from a norm γ and every line l there exists a direction $u \in \mathbb{R}^2$ with $\gamma(u) = 1$ such that

$$d(x, l) = d_u(x, l) \ \forall x \in \mathbb{R}^2,$$

in which case $d_u(x, l) < \infty$. Furthermore, if the unit ball B of γ is strictly convex this direction u is unique, since a strictly convex body touches a line in at most one point. Now let $S \in \mathcal{S}$ be a segment and $l = l^S$ the line passing through S. Define $cover_d(S)$ as in Definition 30. Then

$$d(x, S) = d_u(x, S) \quad \text{for all } x \in cover_d(S)$$

and $cover_{d_u}(S) = cover_d(S)$.

QED

Now we can prove the main results of this section.

Theorem 8.2 *Let d be a distance derived from a norm and let $\epsilon > 0$. Then $f^*_{l_0} > f^*_{l_0+\epsilon}$ for all l_0 with $0 \le l_0 < l_{crit}$.*

Proof: Consider two exhaustive cases:

Case 1: Assume that $\mathcal{E}x \subseteq cover_d(S^*_{l_0})$ and calculate

$$f^*_{l_0} = f(S^*_{l_0}) = \sum_{m \subset \mathcal{M}} d(Ex_m, S^*_{l_0}) = \sum_{m \in \mathcal{M}} d_t(Ex_m, S^*_{l_0}),$$

where $t \in \mathbb{R}^2$ is a fixed direction and its existence is guaranteed by Lemma 8.8. Let D be a transformation with

1. $D(t) = e_2$ and
2. $det(D) = 1$,

e.g., D can be chosen as a rotation. Then D satisfies the conditions of Lemma 3.9 (see page 60) such that for every $x \in \mathbb{R}^2$ and $S \in \mathcal{S}$

$$d_t(x, S) = d_{ver}(D(x), D(S)).$$

Applying D to the set $\mathcal{E}x$ and S^*_l yields that

$$\sum_{m \in \mathcal{M}} d_t(Ex_m, S^*_{l_0}) = \sum_{m \in \mathcal{M}} d_{ver}(D(Ex_m), D(S^*_{l_0})).$$

Choose a segment $S'' \in \mathcal{S}$ with length $le(S'') = l_0 + \epsilon, \epsilon > 0$, such that

$$\sum_{m \in \mathcal{M}} d_{ver}(D(Ex_m), D(S^*_{l_0})) > \sum_{m \in \mathcal{M}} d_{ver}(D(Ex_m), S'')$$

according to Theorem 8.1. Defining $S' = D^{-1}(S'')$ we obtain that $le(S') = le(S'') = l_0 + \epsilon$, and therefore

$$\begin{aligned} f(S^*_{l_0}) &= \sum_{m \in \mathcal{M}} d_t(Ex_m, S^*_{l_0}) \\ &= \sum_{m \in \mathcal{M}} d_{ver}(D(Ex_m), D(S^*_{l_0})) \\ &> \sum_{m \in \mathcal{M}} d_{ver}(D(Ex_m), S'') \\ &= \sum_{m \in \mathcal{M}} d_t(Ex_m, D^{-1}(S'')) \\ &\geq \sum_{m \in \mathcal{M}} d(Ex_m, S') \text{ due to Lemma 8.7} \\ &= f(S') \geq f(S^*_{l_0+\epsilon}). \end{aligned}$$

This implies that $f^*_{l_0} > f^*_{l_0+\epsilon}$.

Case 2: Assume that there exists an $Ex_k \in \mathcal{E}x$ such that $Ex_k \notin cover_d(S^*_{l_0})$. Define a new segment $S' \in \mathcal{S}$ by lengthening an optimal segment $S^*_{l_0}$ towards the facility Ex_k and such that $S^*_{l_0} \subseteq S'$ and $le(S') = l_0 + \epsilon, \epsilon > 0$. Choose a and y such that

$$\begin{aligned} d(Ex_k, S^*_{l_0}) &= d(Ex_k, a), \text{ and} \\ d(Ex_k, l^{S^*_{l_0}}) &= d(Ex_k, y), \; y \in l^{S^*_{l_0}}, \end{aligned}$$

where a is an endpoint of $S^*_{l_0}$. Then, due to the assumption,

$$d(Ex_k, y) < d(Ex_k, a).$$

Since $d(Ex_k, x)$ is convex with respect to x for $x \in l^{S^*_{l_0}}$ and

$$\min_{x \in l^{S^*_{l_0}}} d(Ex_k, x) = d(Ex_k, y) = d(Ex_k, a) < \min_{x \in S^*_{l_0}} d(Ex_k, x),$$

$d(Ex_k, x)$ is strictly convex for $x \in S^*_{l_0}$. Consequently,

$$d(Ex_k, S') < d(Ex_k, S^*_{l_0}).$$

Also for every $m \in \mathcal{M}, m \neq k$, we have $d(Ex_m, S') \leq d(Ex_m, S^*_{l_0})$. Summarizing, we get $f(S^*_{l_0+\epsilon}) \leq f(S') < f(S^*_{l_0})$ yielding $f^*_{l_0} > f^*_{l_0+\epsilon}$.

QED

Theorem 8.3 *Let d be a distance derived from a norm and let $\epsilon > 0$. Then $g^*_{l_0} > g^*_{l_0+\epsilon}$ for all l_0 with $0 \leq l_0 < l_{crit}$.*

Proof: Define the set

$$\bar{\mathcal{M}} := \{k \in \mathcal{M} : Ex_k \notin cover_d(S^*_{l_0})\}$$

and consider the following cases:

Case 1: $\bar{\mathcal{M}} = \emptyset$ which implies that $\mathcal{E}x \subseteq cover_d(S^*_l)$. Then follow the proof of Theorem 8.2, Case 1, to obtain the desired result.

Case 2: $\bar{\mathcal{M}} \neq \emptyset$ and

(a) none of the facilities in $\bar{\mathcal{M}}$ determines $g(S^*_{l_0})$, i.e., for all $k \in \bar{\mathcal{M}}$

$$d(Ex_k, S^*_{l_0}) < \max_{m \in \mathcal{M} \setminus \bar{\mathcal{M}}} d(Ex_m, S^*_{l_0}) = g(S^*_{l_0}),$$

or

(b) there exists at least one facility $Ex_k, k \in \bar{\mathcal{M}}$, that determines $g(S^*_{l_0})$, i.e., there exists a $k \in \bar{\mathcal{M}}$ such that

$$d(Ex_k, S^*_{l_0}) = g(S^*_{l_0}).$$

Now we continue the proof for the cases 2(a) and 2(b).

Case 2(a): : First observe that

$$\begin{aligned} g^*_{l_0} &= \max_{m \in \mathcal{M}} d(Ex_m, S^*_{l_0}) \\ &= \max_{m \in \mathcal{M} \setminus \bar{\mathcal{M}}} d(Ex_m, S^*_{l_0}). \end{aligned}$$

Choose a segment S' of length $l_0 + \epsilon', 0 < \epsilon' < \epsilon$, and such that

$$\max_{m \in \mathcal{M} \setminus \bar{\mathcal{M}}} d(Ex_m, S') < \max_{m \in \mathcal{M} \setminus \bar{\mathcal{M}}} d(Ex_m, S^*_{l_0})$$

according to the proof of Theorem 8.2, Case 1. As ϵ' is sufficiently small, we also have for all $k \in \bar{\mathcal{M}}$

$$d(Ex_k, S') \leq \max_{m \in \mathcal{M} \setminus \bar{\mathcal{M}}} d(Ex_m, S'),$$

which implies that

$$g(S') = \max_{m \in \mathcal{M} \setminus \bar{\mathcal{M}}} d(Ex_m, S').$$

Since lengthening the segment S' to the length $l_0 + \epsilon$ will not increase $g(S')$, we have $g^*_{l_0+\epsilon} = g(S^*_{l_0+\epsilon}) \leq g(S')$. Summarizing, this yields $g^*_{l_0} > g^*_{l_0+\epsilon}$.

Case 2(b): : We consider two subcases:

(1) The optimal value $g^*_{l_0}$ is determined **only** by facilities $Ex_k, k \in \bar{\mathcal{M}}$. Lengthen $S^*_{l_0}$ by ϵ, $\epsilon > 0$, towards these facilities Ex_k, so that we get a segment S' of length $l_0 + \epsilon$. (Remark: If there are two facilities $Ex_k, Ex_r, k, r, \in \bar{\mathcal{M}}$, located in the two disjoint halfplanes generated by $cover_d(S^*_{l_0})$ such that

$$g(S^*_{l_0}) = d(Ex_k, S^*_{l_0}) = d(Ex_r, S^*_{l_0}),$$

then lengthen $S^*_{l_0}$ on both ends by $\frac{\epsilon}{2}$.) Then for all $m \in \mathcal{M}$

$$d(Ex_m, S') \leq g(S^*_{l_0}),$$

and hence $g(S^*_{l_0+\epsilon}) \leq g(S') < g(S^*_{l_0}) = g^*_{l_0}$, which yields $g^*_{l_0} > g^*_{l_0+\epsilon}$.

(2) The optimal value is determined by a facility $Ex_k, k \in \bar{\mathcal{M}}$, **as well as** by at least one facility $Ex_m, m \in \mathcal{M} \setminus \bar{\mathcal{M}}$. Lengthen $S^*_{l_0}$ by $\frac{\epsilon}{2}$ towards the facility Ex_k, so that we get a segment S' of length $l_0 + \frac{\epsilon}{2}$. Then there exists an $Ex_r \in cover_d(S')$ such that

$$d(Ex_r, S') > \max_{Ex_m \notin cover_d(S')} d(Ex_m, S'),$$

which implies that $g(S')$ is determined by a facility in $cover_d(S')$. This brings us to the Case 2(a), the proof of which we adopt, but with $\epsilon' := \frac{\epsilon}{2}$ and obtain the desired result.

QED

Theorems 8.2 and 8.3 can be used to characterize the whole set of efficient line segments in the two bicriteria segment location problems.

Corollary 8.4 *Let d be a distance derived from a strictly convex norm. Then for all $S \in \mathcal{S}$*

$$0 \leq le(S) \leq l_{crit} \text{ and } h(S) = h^*_{le(S)} \Longleftrightarrow S \in Eff_h(\mathcal{S}),$$

i.e., all optimal solutions of P_{l_0} (or of $P_{\leq l_0}$) with $0 \leq l_0 \leq l_{crit}$ are efficient, and vice versa.

There are two cases in which P_{l_0} can be solved efficiently. One is $1S/\mathbb{R}^2/\cdot/d_{ver}/\cdot$ which has been discussed in Section 6.4. The other one is the location of a central line segment with respect to the Euclidean distance $1S/\mathbb{R}^2/\cdot/l_2/\max$, see [ILY92] and [AEST93] for a discussion of structural results and an efficient solution approach. In both cases it is possible to determine efficient line segments with given length.

8.2 Location of a Circle

Another interesting example for the location of one-dimensional objects is the location of a **circle**

$$C = C(c, r) = \{y \in \mathbb{R}^2 : d(c, y) = r\},$$

dependent on three variables: the two coordinates of the **center** $c = (c_1, c_2) \in \mathbb{R}^2$ and the **radius** $r \geq 0$. Assuming that $d = l_2$ is the Euclidean distance, the distance between a point $x \in \mathbb{R}^2$ and a circle $C = C(c, r)$ is given by

$$d(x, C) = \min_{y \in C} d(x, y) = |l_2(x, c) - r|.$$

The formulation of the circle location problem in the plane is as follows. Given a set of existing facilities $\mathcal{E}x$ with positive weights $w_m, m \in \mathcal{M}$, in the plane, we are looking for a circle C dependent on $c \in \mathbb{R}^2$ and on the radius $r \geq 0$ such that we minimize one of the following objective functions.

In the median case:

$$f(C) = \sum_{m \in \mathcal{M}} w_m d(Ex_m, C).$$

In the center case:

$$g(C) = \max_{m \in \mathcal{M}} w_m d(Ex_m, C).$$

Note that both problems are non-differentiable and non-convex optimization problems. In the classification scheme they are given by $1C/\mathbb{R}^2/\cdot/l_2/\sum$ and $1C/\mathbb{R}^2/\cdot/l_2/\max$. Circle location problems have applications in production processes, see e.g. [VY89, YV88]. The center problem is also called *out-of-roundness problem*. In [LL91] it has been shown that in the unweighted case the center c of an optimal circle can be found by investigating the intersection points of the closest and farthest point Voronoi diagram. Heuristic approaches for finding a center circle are given in [DSW96]. The above circle location problems and some related problems have been discussed in [WGHS97], and [Spä97a, Spä97b] deals with circle and ellipse location problems minimizing the sum of *squared* distances to a given point set. Nothing seems to be known about the location of a median circle.

In the following we will present some ideas for solving the circle location problem in the median case. We use the denotation **median circle** for any optimal solution of the median circle location problem. First, consider the case of only $M \leq 3$ existing facilities. For one or two existing facilities there exists an infinite number of circles passing through all existing facilities, such that the problem is easily solved by some circle C with $f(C) = 0$. If $M = 3$ we distinguish two cases.

Case 1 : The three existing facilities are affinely independent. Then there exists exactly one circle $C = C(c, r)$ containing all existing facilities, namely the unique circumcircle of the triangle defined by Ex_1, Ex_2, and Ex_3.

Case 2 : If Ex_1, Ex_2, and Ex_3 are collinear, i.e., they are contained in one common line l, then no optimal circle with finite radius exists, since approximating the line by a infinitely large circle (the center of it moving to infinity) leads to the optimal solution l which can be interpreted as a degenerate circle.

So far it is not clear whether this result for only three existing facilities can be extended to the case $M > 3$. Since for line location problems $1l/\mathbb{R}^2/\cdot/\gamma/\sum$ there always exists a median line passing through two of the existing facilities, one might suspect that in the circle location problem there always exists a median circle which contains three of the existing facilities. For $M = 3$ this is true according to the case analysis above. But the following example shows that the statement *there always exists an optimal circle which contains three of the existing facilities* is wrong for general M.

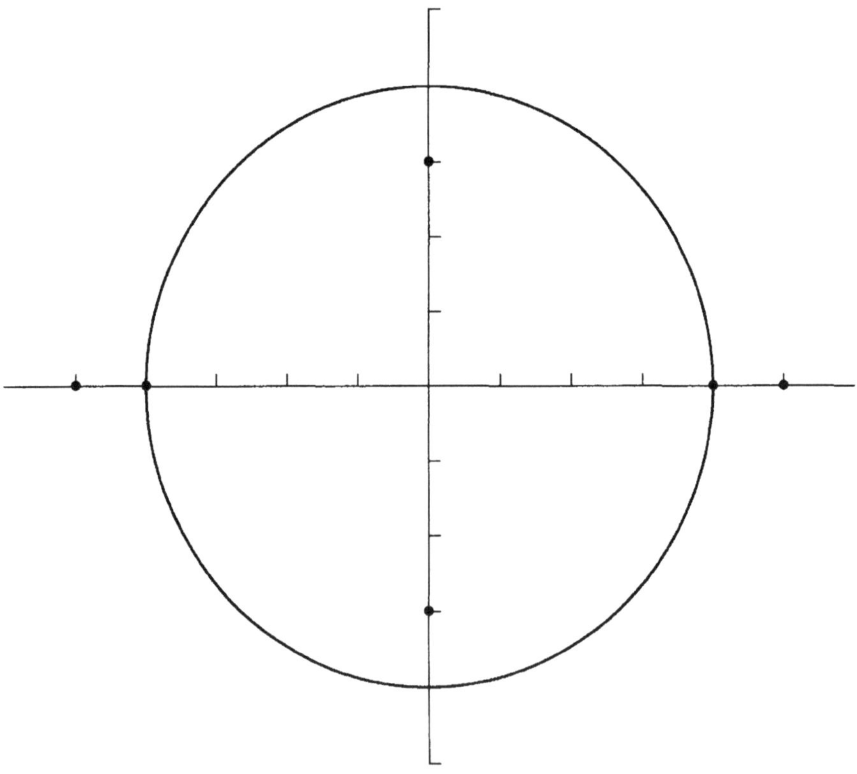

Figure 8.1. An example in which no median circle contains three of the existing facilities.

Example 8.1 ([Pee96]) *Consider the following set of $M = 6$ existing facilities with equal weights $w_m = 1, m = 1, \ldots, 6$ (see Figure 8.1):*

$$\begin{aligned} Ex_1 &= (-4, 0) \\ Ex_2 &= (0, 3) \\ Ex_3 &= (0, -3) \\ Ex_4 &= (4, 0) \\ Ex_5 &= (5, 0) \\ Ex_6 &= (-5, 0) \end{aligned}$$

In this example, the circle $C^* = C(0, 4)$ with center $(0, 0)$ and radius $r = 4$ leads to

$$f(C^*) = 4$$

and is therefore better than all circles passing through three of the existing facilities. (The best of these circles are the circles C_{124} and C_{134} passing through the existing facilities Ex_1, Ex_2, and Ex_4, and through the existing facilities Ex_1, Ex_3, and Ex_4, respectively. Both lead to $f(C_{124}) = f(C_{134}) \approx 4.269$.)

Although Example 8.1 shows that there may exist only optimal circles that do not pass through three of the existing facilities, numerical tests indicate that in most cases (at least in small examples) there exists an optimal solution containing three of the existing facilities, i.e., c is the center of the circumcircle of a triangle defined by three of the existing facilities. Note that even in the counterexample, c lies on the intersection point of two bisectors of existing facilities.

Lemma 8.9 *There always exists an optimal circle C of the median circle location problem that contains one of the existing facilities.*

Proof: Suppose that $C = C(c, r)$ is an optimal circle that does not contain any of the existing facilities. Consider the following one-dimensional function, where c has been fixed and we only look at the radius r.

$$\bar{f}(r) = f(C(c, r)) = \sum_{m \in \mathcal{M}} w_m |c_m - r|,$$

where $c_m = l_2(Ex_m, c)$ is the Euclidean distance between Ex_m and the fixed center c of the circle. According to Lemma 1.9 on page 22 we conclude that there exists a $k \in \mathcal{M}$ such that $r^* = c_k$ is a minimizer of $\bar{f}$. The corresponding circle $C^* = C(c, r^*)$ then satisfies

$$f(C^*) = \bar{f}(r^*) \leq \bar{f}(r) = f(C).$$

Hence C^* is also optimal for the original circle location problem and passes through Ex_k.

QED

Definition 32 *A circle $C = C(c, r)$ is called* **pseudo-halving** *if the following holds.*

$$\sum_{m:d(Ex_m,c)<r} w_m \leq \frac{W}{2} \text{ and}$$

$$\sum_{m:d(Ex_m,c)>r} w_m \leq \frac{W}{2}.$$

Lemma 8.10 *All median circles are pseudo-halving.*

Proof: Let $c \in \mathbb{R}^2$ be an arbitrary center point. Consider the same one-dimensional function $\bar{f}$ as in the proof of Lemma 8.9. From Lemma 1.9 and Definition 13 we know that the set of minimizers of $\bar{f}$ is given by

$$median\{(c_m, w_m) : m \in \mathcal{M}\} = \left\{ r : \sum_{m:c_m<r} w_m \leq \frac{W}{2} \text{ and } \sum_{m:c_m>r} w_m \leq \frac{W}{2} \right\}.$$

Hence all optimal circles with center c are pseudo-halving. Since this holds for all $c \in \mathbb{R}^2$, the result follows.

QED

In the following we give two possible interpretations of circle location problems. The first reduces the circle location problem to a point-shaped facility location problem in the plane with positive and negative weights and a given feasible polyhedral set, and the other interpretation uses a dual transformation to obtain a three-dimensional point-shaped location problem with the vertical distance, but where the set of existing facilities is given by cones.

Reduction to $1/\mathbb{R}^2/R, w_m \in \mathbb{R}/l_2/\sum$ Because of Lemma 8.9 we can assume that one of the existing facilities lies on an optimal circle C^*. Hence for all existing facilities Ex_m we can solve the following problem P_m:

$$\min f(C)$$

$$s.t.\ Ex_m \in C.$$

Without loss of generality assume that $Ex_k = 0$. Then the objective function of P_k can be rewritten as follows.

$$\begin{aligned} f(C) &= \sum_{m\in\mathcal{M}} w_m |l_2(Ex_m, c) - r| \\ &= \sum_{m\in\mathcal{M}} w_m |l_2(Ex_m, c) - l_2(c, 0)|. \end{aligned}$$

To get rid of the absolute values we introduce

$$\begin{aligned} \mathcal{M}^{in}(c) &= \{m \neq k, m \in \mathcal{M} : l_2(c, Ex_m) \leq l_2(c, 0)\} \text{ and} \\ \mathcal{M}^{out}(c) &= \{m \in \mathcal{M} : l_2(c, Ex_m) > l_2(c, 0)\}. \end{aligned}$$

Defining

$$w'_m = \begin{cases} -w_m & \text{if } m \in \mathcal{M}^{in}(c) \\ w_m & \text{if } m \in \mathcal{M}^{out}(c) \\ \sum_{m\in\mathcal{M}^{in}(c)} w_m - \sum_{m\in\mathcal{M}^{out}(c)} w_m & \text{if } m = k, \end{cases}$$

we obtain the following objective function.

$$\begin{aligned} f(c)) &= \sum_{m\in\mathcal{M}^{in}(c)} w_m(l_2(c,0) - l_2(c,Ex_m)) \\ &\quad + \sum_{m\in\mathcal{M}^{out}(c)} w_m(l_2(c,Ex_m) - l_2(c,0)) \\ &= \sum_{m\in\mathcal{M}\setminus\{k\}} w'_m(l_2(c,Ex_m) - l_2(c,0)) \\ &= \sum_{m\in\mathcal{M}} w'_m l_2(c,Ex_m). \end{aligned}$$

This is the objective function of the classical Weber problem with positive and negative weights. Additionally, in our case

$$\sum_{m\in\mathcal{M}} w'_m = 0.$$

The two sets $\mathcal{M}^{in}(c)$ and $\mathcal{M}^{out}(c)$ are obviously dependent on the center c of the circle we are looking for. In order to solve P_k we therefore have to investigate all feasible partitions of $\mathcal{M} \setminus \{k\}$ into sets $\mathcal{M}^{in}$ and $\mathcal{M}^{out}$. Such a partition $\{\mathcal{M}^{in}, \mathcal{M}^{out}\}$ is feasible if there exists a $c \in \mathbb{R}^2$ such that

$$\begin{aligned} \mathcal{M}^{in}(c) &= \mathcal{M}^{in} \text{and} \\ \mathcal{M}^{out}(c) &= \mathcal{M}^{out}. \end{aligned}$$

Additionally, we can use Lemma 8.10 and consider only partitions that belong to pseudo-halving circles, i.e. that satisfy

$$\begin{aligned} \sum_{m\in\mathcal{M}^{in}} w_m &\le \frac{W}{2} \text{ and} \\ \sum_{m\in\mathcal{M}^{out}} w_m &\le \frac{W}{2}. \end{aligned}$$

Summarizing, we have to solve the following location problem for all feasible partitions of $\mathcal{M} \setminus \{k\}$.

$$\min f(c)$$

$$\begin{aligned} s.t. \quad & l_2(c,Ex_m) \le l_2(c,0) \quad \text{for all } m \in \mathcal{M}^i \\ & l_2(c,Ex_m) \ge l_2(c,0) \quad \text{for all } m \in \mathcal{M}^o, \end{aligned}$$

where $\mathcal{E}x$ is the set of existing facilities with positive and negative weights w'_m, given by

$$w'_m = \begin{cases} -w_m & \text{if } m \in \mathcal{M}^i(c) \\ w_m & \text{if } m \in \mathcal{M}^o(c) \\ \sum_{m \in \mathcal{M}^{in}(c)} w_m - \sum_{m \in \mathcal{M}^{out}(c)} w_m & \text{if } m = k. \end{cases}$$

This approach can be used to solve the circle location problem algorithmically by inspecting all problems $P_m, m \in \mathcal{M}$. To solve the problem P_m, a branch and bound approach can be used together with a barrier function modelling the feasible region. Details of such an approach are under consideration, see [FS98].

Reduction to $1/\mathbb{R}^3/\mathcal{E}x = \{cones\}/d_{ver}/\sum$ Here we introduce a transformation T^c which maps points to 3-dimensional cones and circles to 3-dimensional points.

Definition 33 *We define the following* **transformation T^c**.
For any point $x = (x_1, x_2) \in \mathbb{R}^2$ define

$$T^c(x) \quad := \quad \{(y_1, y_2, y_3) \in \mathbb{R}^3 : y_3 \geq 0 \text{ and } l_2(x, (y_1, y_2)) = y_3\}$$

as the boundary of a cone, and for any circle $C = C(c, r)$ with center $c = (c_1, c_2)$ and radius r define $T^c(C)$ as a point

$$T^c(C) \quad := \quad (c_1, c_2, r) \in \mathbb{R}^3.$$

The lower envelope of the set $\{T^c(Ex_m) : m \in \mathcal{M}\}$ is also called the Voronoi surface of $\mathcal{E}x$ (see [AG96]). This Voronoi surface can be defined as the graph of the function

$$d_{\min}(x) = \min_{m \in \mathcal{M}} d(x, Ex_m).$$

The transformation T^c transforms Euclidean distances to vertical distances as the following lemma shows.

Lemma 8.11 *Let C be a circle and $x \in \mathbb{R}^2$ be a point.*

1. $l_2(x, C) = d_{ver}(T^c(x), T^c(C))$
2. $l_2(x, C) = l_2(T^c(x) \cap H, T^c(C))$, *where* $H = \{(y_1, y_2, y_3) \in \mathbb{R}^3 : y_3 = r\}$.

Proof: Using the definition of $T^c(x)$ we get

$$\begin{aligned} l_2(x, C) &= |l_2(x, c) - r| \\ &= |d_{ver}((c_1, c_2, 0), (c_1, c_2, l_2(x, c))) - r| \\ &= |d_{ver}((c_1, c_2, 0), T^c(x)) - r| \\ &= d_{ver}((c_1, c_2, r), T^c(x)) \\ &= d_{ver}(T^c(C), T^c(x)). \end{aligned}$$

To see also the second part of the lemma we use the definition of $T^c(x)$ and note that for any point $y = (y_1, y_2, r) \in \mathbb{R}^3$

$$\begin{aligned} d_{ver}(T^c(x), y) &= |l_2(x, (y_1, y_2)) - r| \\ &= |l_2((x_1, x_2, r), (y_1, y_2, r)) - r| \\ &= l_2((y_1, y_2), C(x, r)) \\ &= l_2(T^c(x) \cap H, T^c(C)), \end{aligned}$$

since $T^c(x) \cap H$ is a circle with center (x_1, x_2, r) and radius r.

QED

Lemma 8.11 allows two interpretations of the circle location problem. Part 1 shows that it is equivalent to locating a point $x \in \mathbb{R}^3$. The objective of this 3-dimensional location problem is to minimize the sum of vertical distances to a given set of cones $\{T^c(Ex_m) : m \in \mathcal{M}\}$. To deal with this formulation, it might be helpful to look at a vertical line $l = l(c_1, c_2, 0)$ passing through the point $(c_1, c_2, 0)$, meaning that we fix the center point $c = (c_1, c_2)$ of the circle. Then it is easy to determine the best radius r (see Lemma 8.9), i.e., we solve a simple Weber problem on a straight line with $\mathcal{E}x_W = \{l \cap T^c(Ex_m) : m \in \mathcal{M}\}$ as the set of existing facilities. While translating the line l to another vertical line the set of existing facilities $\mathcal{E}x_W$ of the simple Weber problem changes continuously. Coincidence of two facilities $l \cap T^c(Ex_m) = l \cap T^c(Ex_k)$ in $\mathcal{E}x_W$ means that $c = (c_1, c_2)$ lies on the bisector of Ex_m and Ex_k.

The second part of Lemma 8.11 deals with horizontal hyperplanes

$$H_r = \{y = (y_1, y_2, r) \in \mathbb{R}^3\}.$$

On each hyperplane H_r we have a given set of circles $\{T^c(Ex_m) \cap H : m \in \mathcal{M}\}$ as existing facilities and we want to locate a point $x \in H$ such that the sum of distances from x to the circles is minimized. Since all circles have the same radius r we can use the equation

$$l_2(x, C(y, r)) = l_2(C(x, r), y)$$

to give the following equivalent formulation of this problem in the plane: Given a set of existing (point-shaped) facilities $\mathcal{E}x$ in the plane, find a circle C with fixed radius r such that

$$f(C) = \sum_{m \in \mathcal{M}} w_m l_2(Ex_m, C)$$

is minimized.

8.3 Planar Location of One-Dimensional Facilities

From a practical point of view it would be nice to be able to locate curves of arbitrary shape in the plane. Bus lines, highways, pipelines, and conveyor belts may sometimes be assumed to be linear, but in real life they can be of any shape. In the following we formulate the decision version of such a general location problem. We define the length $le(New)$ of a one-dimensional facility as its Euclidean length.

(GLP) General Location Problem

Instance : A finite set $\mathcal{E}x = \{Ex_m : m \in \mathcal{M}\}$ of existing facilities $Ex_m \in \mathbb{R}^2$ and $h_0 \geq 0$, $l_0 \geq 0$, $h_0, l_0 \in \mathbb{R} \cup \{\infty\}$.

Question : Does there exist a facility *New* (a path, a closed path, a tree, a circle, ...) such that the length $le(New) \leq l_0$ and the distance to the existing facilities $h(New) \leq h_0$, where $h \in \{f, g\}$?

Note that if $h_0 = 0$ then (GLP) with $h = f$ is equivalent to (GLP) with $h = g$. The following problems are special versions of (GLP).

- If l_0=0, *New* is forced to be a point and we have a classical facility location problem in the plane, for $h = f$ we get the Weber problem $1/\mathbb{R}^2/\cdot/d/\sum$ and for $h = g$ we have the center problem $1/\mathbb{R}^2/\cdot/d/\max$.
- If *New* is a line segment, the above formulation is the decision version of the bicriteria segment location problem of Section 8.1.
- If *New* is a line segment and $l_0 = \infty$, then (GLP) describes line location problems.
- If *New* consists of r lines and $h_0 = 0$, $l_0 = \infty$, we have the line cover problem of [MT82]. In Section 3.2 we already mentioned that this problem is NP-hard.
- If *New* is a circle and $l_0 < \infty$, (GLP) is the problem of locating a circle with bounded radius $r \leq \frac{l_0}{2\Pi}$.
- If *New* is a circle and $l_0 = \infty$, we face the circle location problem of Section 8.2.
- If *New* is a closed path and $h_0 = 0$, then (GLP) is a formulation of the geometric traveling salesman problem (see, e.g., [GJ79], problem [ND23]).
- If *New* is a closed path, $h_0 > 0$ and $h = g$, then we get a special case of the geometric covering salesman problem. In this problem one looks for a tour of minimum length that visits a set of neighbourhoods specified by the clients. This problem can be reduced to the geometric traveling salesman problem. In [AH94] approximation algorithms for three versions of the problem are given:
  - The neighbourhoods are parallel line segments; this corresponds to $d = d_t$.

- The neighbourhoods are translates of a polytope P. In this case the distance d has to be derived from a polyhedral gauge $\tilde{\gamma}$ whose unit ball equals P.
- The neighbourhoods are circles. This case is modelled by $d = l_2$ in the above formulation of (GLP).

- If *New* is a path and $h_0 = 0$, (GLP) is a Hamiltonian path problem in the plane.
- If *New* is a tree and $h_0 = 0$, the resulting problem is the geometric Steiner tree problem, (see, e.g., [GJ79], problem [ND13]).

Lemma 8.12 *If New is a tree, a path , or a closed path, the general location problem (GLP) is NP-hard.*

Proof: If *New* is a tree, the geometric Steiner tree problem is a special case of (GLP), and if *New* is a closed path, (GLP) contains the geometric traveling salesman problem as a special case. Since both problems are NP-hard for $d = l_2$, $d = l_1$ and $d = l_\infty$ (see [GJ79], problems [ND13] and [ND23]), (GLP) is also NP-hard in these cases. To show that (GLP) is NP-hard in the case that *New* is a non-closed path we use exercise 13(b) in [JP85]. According to this, it is NP-hard to decide whether there exists a spanning tree for a set of points in the Euclidean plane which is a path. Now suppose we have a given set of points $\mathcal{E}x$ in the Euclidean plane. We then determine a minimal spanning tree T of $\mathcal{E}x$ in polynomial time (see e.g. [BM76]). We then solve (GLP) in the case that *New* is a path, $d = l_2$, and $h_0 = 0$. Let P be a solution path with length $le(P)$. If $le(P) = le(T)$, then there exists a minimal spanning tree which is a path, and if $le(P) > le(T)$, then such a tree does not exist. This reduction shows that (GLP) is also an NP-hard problem in the case that *New* is a path.

QED

If *New* is a path or a closed path in the plane, the following lemma shows that if the answer to (GLP) is YES then there exists a solution of (GLP) which is a polygonal path and a polygonal closed path, respectively. This means, when solving a corresponding minimization problem, it suffices to investigate all polygonal paths (or polygonal closed paths) to find an optimal solution.

Lemma 8.13 *Consider (GLP) where d is derived from a norm γ, New is a (closed) path and l_0, h_0 are given. Let New* be a solution of (GLP), i.e., $le(New^*) \leq l_0$ and $h(New^*) \leq h_0$. Then there exists a polygonal (closed) path New′ such that*

$$\begin{aligned} le(New') &\leq le(New^*) \leq l_0 \text{ and} \\ h(New') &\leq h(New^*) \leq h_0. \end{aligned}$$

In the case that New is a non-closed path, New′ has at most $M-2$ breakpoints; in the case that New is a closed path we get at most $M-1$ breakpoints for New′.

Proof: For each $m \in \mathcal{M}$ determine a point $x_m \in New^*$ such that

$$d(Ex_m, New^*) = d(Ex_m, x_m).$$

Then $x_m \in New^*$ for all $m \in \mathcal{M}$. Without loss of generality assume that the x_m are ordered in such a way that x_m and x_{m+1}, $m = 1, \ldots, M-1$, are directly connected by a curve segment of New^*. (If New^* is a closed path then additionally x_1 and x_M are connected.) Replacing all these curve segments by straight lines gives us a polygonal (closed) path New' with at most $M-2$ breakpoints $x_2, \ldots, x_{M-1}$ if New' is a path, and with at most $M-1$ breakpoints $x_M = x_1, x_2, \ldots, x_{m-1}$ if New' is a closed path. Furthermore, the Euclidean length of the line segment from x_m to x_{m+1} is never larger than the length of any other curve connecting x_m and x_{m+1}, and therefore $le(New') \leq le(New^*)$. Finally, we note that

$$d(Ex_m, New') \leq d(Ex_m, x_m) = d(Ex_m, New^*)$$

for all $m \in \mathcal{M}$ and hence $h(New') \leq h(New^*)$.

QED

Appendix A
Summary

The two tables on the following page provide a concise overview of the main properties of line and hyperplane location problems as discussed in this book.

Med0	There exists a median hyperplane passing through one existing facility.
Med1	There exists a median hyperplane passing through n affinely independent existing facilities.
Med2	Every median hyperplane is pseudo-halving.
Cen1	There exists a center hyperplane at maximum (weighted) distance from $n+1$ of the existing facilities.
Cen1'	There exists a center hyperplane at maximum (weighted) distance from $n+1$ affinely independent existing facilities.
Cen2	If the weights are all equal, there exists a center hyperplane parallel to a facet of the convex hull of the existing facilities.

	Med0	Med1	Med2	Cen1	Cen1'	Cen2
In the plane						
t-distances	yes	yes	yes	yes	yes	yes
norms	yes	yes	yes	yes	yes	yes
gauges	yes	no	no	yes	*yes*[1]	yes
strictly monotone metrics	no	no	no	yes	*no*[1]	no
metrics	no	no	no	no	no	no
mixed gauges	yes	no	no	yes	*no*[1]	no
In $\mathbb{R}^n$						
t-distances	yes	yes	yes	yes	yes	no
norms	yes	yes	yes	yes	yes	no
gauges	*yes*[1]	no	no	*yes*[1]	*yes*[1]	no
strictly monotone metrics	no	no	no	*yes*[1]	*no*[1]	no
metrics	no	no	no	no	no	no
mixed gauges	*yes*[1]	no	no	*yes*[1]	*no*[1]	no

1) without proof

Appendix B
List of Algorithms

Appendix C
List of Symbols

General Notation

Let $A, B \subseteq \mathbb{R}^n$.

$conv(A)$	convex hull of A
$aff(A)$	affine hull of A
$int(A)$	interior of A
∂A	boundary of A
$Ext(A)$	extreme points of A
$\|A\|$	number of elements in A
$dim(A)$	dimension of A
$b_{A,B}$	bisector of A and B
$b^w_{A,B}$	weighted bisector of A and B

Location Theory

M	number of existing facilities
$\mathcal{M}$	index set of existing facilities
Ex_m	existing facility
$\mathcal{E}x$	set of existing facilities
w_m	weight of existing facility Ex_m
W	sum of all weights
f	sum objective function
g	center objective function
h	refers to both f and g independently
New	a new facility
l	a line
s	slope of a line
S	a line segment
H	a hyperplane
$\mathfrak{n}$	normal vector of a hyperplane

C	a circle
z^*	objective value of the location problem
l^*	an optimal line
H^*	an optimal hyperplane
$\mathcal{L}^*$	set of all optimal lines
R	restricting set
z^*_R	objective value of restricted location problem
$\mathcal{L}^*_R$	set of all optimal lines of restricted problem
$\mathcal{B}^-_H, \mathcal{B}^+_H$	the two open halfspaces separated by the hyperplane H

Distance Measures

d	general distance measure
d_{ver}	vertical distance
d_{hor}	horizontal distance
d_t	t-distance
l_1	rectangular norm
l_2	Euclidean norm
$\|t\|$	Euclidean length of $t \in \mathbb{R}^n$
l_∞	Chebyshev norm
l_p	p-norm
γ	arbitrary norm
γ_B	block norm
$\tilde{\gamma}$	arbitrary gauge
$\tilde{\gamma}_B$	polyhedral gauge
B	unit ball of a norm or a gauge
$\mathcal{CP}_\gamma$	corner points of the norm γ

Piecewise Linear Programs

$\mathcal{C}$	set of all cells
$\mathcal{C}^\partial$	set of all facets of all cells
$\mathcal{H}$	set of construction lines
$\mathcal{P}(\mathcal{H})$	points on construction lines
$\mathcal{K}^{med}$	construction lines of the median line location problem
$\mathcal{K}^{cen}$	construction lines of the unweighted center line location problem
$\mathcal{K}^{cen}_{weights}$	construction lines of the weighted center line location problem
R	restricting set
t^*	objective value
t^*_R	objective value of restricted problem
$\mathcal{X}^*$	set of optimal solutions
$\mathcal{X}^*_R$	set of optimal solutions of the restricted problem
$Cand$	candidate set

References

[AD88] D. Avis and M. Doskas. Algorithms for high dimensional stabbing problems. In *Theoretical Foundations of Computer Graphics and CAD*, volume F40 of *NATO ASI*, pages 199–210. Springer, 1988.

[AEST93] P.K. Agarwal, A. Efrat, M. Sharir, and S. Toledo. Computing a segment center for a planar point set. *Journal of Algorithms*, 15:314–323, 1993.

[AG96] H. Alt and L.J. Guibas. Discrete geometric shapes: Matching, interpolation, and approximation. Technical Report B 96-11, Freie Universität Berlin — Serie B Informatik, 1996.

[AH94] E.M. Arkin and R. Hassin. Approximation algorithms for the geometric covering salesman problem. *Discrete Applied Mathematics*, 55:197–218, 1994.

[ARW89] D. Avis, J.-M. Robert, and R. Wenger. Lower bounds for line stabbing. *Information Processing Letters*, 33:59–62, 1989.

[Bar97] I. Bartling. Standortplanung — Algorithmen zur Positionierung von Geraden in der Ebene. Master's thesis, Universität Kaiserslautern, 1997. Staatsexamensarbeit.

[BFP+72] M. Blum, R.W. Floyd, V.R. Pratt, R.L. Rivest, and R.E. Tarjan. Time bounds for selection. *J. Comput. System Sci.*, 7:448–461, 1972.

[BL93] J. Brimberg and R.F. Love. Directional bias of the l_p norm. *European Journal of Operational Research*, 67(2):287–294, 1993.

[BM76] J.A. Bondy and U.S.R. Murty. *Graph Theory with Applications*. North-Holland, New York-Amsterdam-Oxford, 1976.

[Bos57] R.J. Boscovich. De litteraria expeditione per pontificiam ditionem, et synopsis amplioris operis, ac habentur plura ejus ex exemplaria etiam sensorum impressa. *Bononiensi Scientiarum et Artum Instituto atque Academia Commentarii*, 4:353–396, 1757.

[CC61] A. Charnes and W.W. Cooper. *Management Models and Industrial Applications of Linear Programming*, volume 1. Wiley, New York, 1961.

[CCF55] A. Charnes, W.W. Cooper, and R.O. Ferguson. Optimal estimation of executive compensation by linear regression. *Management Science*, 1:138–151, 1955.

[CGP+90] S.E. Cappell, J.E. Goodman, J. Pach, R. Pollack, M. Sharir, and R. Wenger. The combinatorial complexity of hyperplane transversals. In *Proceedings of the 6th annual symposium on computational geometry*, pages 83–91, 1990.

[CP95] E. Carrizosa and F. Plastria. The determination of a "least quantile of squares regression line" for all quantiles. *Computational Statistics & Data Analysis*, 20:467–479, 1995.

[CRC87] John R. Current, Charles S. Revelle, and Jared L. Cohon. The median shortest path problem: a multiobjective approach to analyze cost vs. accessibility in the design of transportation networks. *Transportation Science*, 21(3):188–197, 1987.

[DBM96] J.M. Diaz-Banez and J.A. Mesa. An algorithm for a rectilinear center trajectory. Technical report, Universidad de Sevilla, 1996. submitted.

[DE94] T.K. Dey and H. Edelsbrunner. Counting triangle crossings and halving planes. *Discrete Computational Geometry*, 12:281–289, 1994.

[Dor84] N.N. Doroshko. The solution of two mainline problems in the plane, 1984. Manuscript deposited in VINITY USSR, 14.06.1984, Reg.no. 39614, 15 pages.

[Dre82] Z. Drezner. On minimax optimization problems. *Mathematical programming*, 22(2):227–230, 1982.

[DSW96] Z. Drezner, S. Steiner, and G.O. Wesolowsky. On the circle closest to a set of points. Technical report, California State University, Department of Management Science and Information Systems, 1996.

[Ede85] H. Edelsbrunner. Finding transversals for sets of simple geometric figures. *Theoretical Computer Science*, 35:55–69, 1985.

[Ede87] H. Edelsbrunner. *Algorithms in Combinatorial Geometry*. Springer, Berlin Heidelberg New York, 1987.

[Edg88] F.Y. Edgeworth. On a new method of reducing observations relating to several quantities. *Phil. Magazine (Series 5)*, 25:184–191, 1888.

[Ehr97] M. Ehrgott. *Multiple Criteria Optimization.* Shaker Verlag, Aachen, 1997.

[ELSS73] P. Erdös, L. Lovász, A. Simmons, and E. Strauss. Dissection graphs of planar point sets. In *A Survey of Combinatorial Theory*, pages 139–149. J. Srivastara et al., 1973.

[Erk96] E. Erkut. Finding line segments. Talk held in Los Angeles, 1996.

[EW89] P. Egyed and R. Wenger. Stabbing pairwise disjoint translates in linear time. In *Proceedings of the 5th Annual Symposium on Computational Geometry*, pages 364–369, 1989.

[Fis61] W.D. Fisher. A note on curve fitting with minimum deviations by linear programming. *Journal Amer. Statist. Association*, 56:359–362, 1961.

[Fli97] J. Fliege. *Effiziente Dimensionsreduktion in Multilokationsproblemen.* Shaker Verlag, Aachen, 1997.

[FMW92] R.L Francis, L.F. McGinnis, and J.A. White. *Facility Layout and Location: An Analytical Approach.* Prentice Hall, New York, 2. edition, 1992.

[For83] O. Forster. *Analysis 1. Differential- und Integralrechnung einer Veränderlichen*, volume 24 of *Vieweg Studium.* Vieweg, Braunschweig, 1983.

[FS98] J. Fliege and A. Schöbel. The median circle problem. working paper, 1998.

[GJ79] M.R. Garey and D.S. Johnson. *Computers and Intractability — A Guide to the Theory of NP-Completeness.* H.W. Freeman and Company, San Francisco, 1979.

[GPW93] J.E. Goodman, R. Pollack, and R. Wenger. Geometric transversal theory. In J. Pach, editor, *New Trends in Discrete and Computational Geometry*, chapter 7, pages 163–198. Springer-Verlag, New York, Inc., 1993.

[Ham95] H.W. Hamacher. *Mathematische Lösungsverfahren für planare Standortprobleme.* Vieweg, Braunschweig, 1995.

[Hay81] W.L. Hays. *Statistics.* Holt, Rinehart and Winston, 3. edition, 1981.

[HC83] Y.Y. Haimes and V. Chankong. *Multiobjective Decision Making — Theory and Methodology.* North Holland, New York, 1983.

[Hel23] E. Helly. Über Mengen konvexer Körper mit gemeinschaftlichen Punkten. *Jahrbuch der Deutschen Mathematiker Vereinigung,* 32:175–176, 1923.

[Her89] J. Hershberger. Finding the upper envelope of n line segments in $O(n \log n)$ time. *Information Processing Letters,* 33:169–174, 1989.

[HII+93] M.E. Houle, H. Imai, K. Imai, J.-M. Robert, and P. Yamamoto. Orthogonal weighted linear L_1 and L_∞-approximation and applications. *Discrete Applied Mathematics,* 43:217–232, 1993.

[HIIR89] M.E. Houle, H. Imai, K. Imai, and J.-M. Robert. Weighted orthogonal linear L_∞-approximation and applications. *Lecture Notes Computer Science,* 382:183–191, 1989.

[HN94] H.W. Hamacher and S. Nickel. Combinatorial algorithms for some 1-facility median problems in the plane. *European Journal of Operational Research,* 79:340–351, 1994.

[HN95] H.W. Hamacher and S. Nickel. Restricted planar location problems and applications. *Naval Research Logistics,* 42:967–992, 1995.

[HN96] H.W. Hamacher and S. Nickel. Multicriteria planar location problems. *European Journal of Operational Research,* 94:66–86, 1996.

[HNS96] H. W. Hamacher, S. Nickel, and A. Schneider. Classification of location problems. Technical Report 19, University of Kaiserslautern, Wirtschaftsmathematik, 1996.

[HSL93] S.L. Hakimi, E.F. Schmeichel, and Martine Labbe. On locating path- or tree-shaped facilities on networks. *Networks,* 23:543–555, 1993.

[HT85] M.E. Houle and G.T. Toussaint. Computing the width of a set. In *Proceedings of the 1st ACM Symposium on Computational Geometry,* pages 1–7, 1985.

[HT96] R. Horst and H. Tuy. *Global Optimization.* Springer, 1996. 3rd edition.

[ILY92] H. Imai, D.T. Lee, and C.-D. Yang. 1-segment center problems. *ORSA Journal on Computing,* 4(4):426–434, 1992.

[JP85] D.S. Johnson and C.H. Papadimitriou. Computational complexity. In E.L. Lawler, J.K. Lenstra, A.H.G. Rinnooy Kan, and D.B. Shmoys, editors, *The Traveling Salesman Problem — A Guided*

Tour of Combinatorial Optimization, chapter 3. Wiley, Chichester, 1985.

[KA97] R. V. Kasturi and P. K. Agarwal. Linear approximation of simple objects. *Information Processing Letters*, 62:89–94, 1997.

[Kar58] O.J. Karst. Linear curve fitting using least deviations. *Journal Amer. Statist. Association*, 53:118–132, 1958.

[KK91] T. Koshizuka and O. Kurita. Approximate formulas of average distances associated with regions and their application to location problems. *Mathematical Programming*, 52:99–123, 1991.

[Kle97] Rolf Klein. *Algorithmische Geometrie*. Addison-Wesley, 1997.

[KM90] N.M. Korneenko and H. Martini. Approximating finite weighted point sets by hyperplanes. *Lecture Notes Computer Science*, 447:276–286, 1990.

[KM93] N.M. Korneenko and H. Martini. Hyperplane approximation and related topics. In J. Pach, editor, *New Trends in Discrete and Computational Geometry*, chapter 6, pages 135–162. Springer-Verlag, New York, Inc., 1993.

[Kor89] N.M. Korneenko. Optimal lines in the plane. *Math. Research*, 51:43–51, 1989.

[Kre91] U. Krengel. *Einführung in die Wahrscheinlichleitstheorie und Statistik*, volume 59 of *Vieweg Studium*. Vieweg, Braunschweig, 1991.

[LC85] D.T. Lee and Y.T. Ching. The power of geometric duality revisited. *Inform. Process. Letters*, 21:117–122, 1985.

[Lee86] D.T. Lee. Geometric location problems and their complexity. *Lecture Notes Computer Science*, 233:154–167, 1986.

[LL91] V.-B. Le and D.T. Lee. Out-of-roundness problem revisited. *IEEE Transactions on Pattern Analysis and Machine Intelligence*, 13:217–223, 1991.

[LMW88] R.F. Love, J.G. Morris, and G.O. Wesolowsky. *Facilities Location*, chapter 3.3, pages 51–60. North-Holland, Amsterdam, 1988.

[Lov71] L. Lovász. On the number of halving lines. In *Ann. Univ. Eötvös Loránd*, number 14 in Sekt. Math., pages 107–108. Budapest, 1971.

[Lov72] R.F. Love. A computational procedure for optimally locating a facility with respect to several rectangular regions. *Journal of Regional Science*, 12:233–242, 1972.

[LW86] D.T. Lee and Y.F. Wu. Geometric complexity of some location problems. *Algorithmica*, 1:193–211, 1986.

[Mar87] H. Martini. Some results and problems around zonotopes. In K. Böröczky and G. Fejes Tóth, editors, *Intuitive Geometry*, number 48 in Coll.Math.Soc. J. Bolyai, pages 383–418. North-Holland, 1987.

[MB96] J.A. Mesa and T.B. Boffey. A review of extensive facility location in networks. *European Journal of Operational Research*, 95(3):592–603, 1996.

[Meg83] N. Megiddo. Linear time algorithms for linear programming in $\mathbb{R}^3$ and related problems. *SIAM Journal on Computing*, 12:759–776, 1983.

[Meg84] N. Megiddo. Linear programming in linear time when the dimension is fixed. *Journal of the Association for Computing Machinery (ACM)*, 31:114–127, 1984.

[Mes95] J.A. Mesa. Continuous location of dimensional structures. Working paper, presented at the Workshop on Locational Analysis, 1995.

[Min67] H. Minkowski. *Gesammelte Abhandlungen, Band 2.* Chelsea Publishing Company, New York, 1967.

[MN80] J.G. Morris and J.P. Norback. A simple approach to linear facility location. *Transportation Science*, 14(1):1–8, 1980.

[MN83] J.G. Morris and J.P. Norback. Linear facility location - solving extensions of the basic problem. *European Journal of Operational Research*, 12:90–94, 1983.

[MS97] H. Martini and A. Schöbel. The minisum hyperplane problem in Minkowski spaces. Technical Report 18, University of Kaiserslautern, Wirtschaftsmathematik, 1997. submitted.

[MS98a] H. Martini and A. Schöbel. Central hyperplanes in normed spaces. Technical report, University of Kaiserslautern, Wirtschaftsmathematik, 1998. in preparation.

[MS98b] H. Martini and A. Schöbel. Median hyperplanes in normed spaces — a survey. Technical Report 36, University of Kaiserslautern, Wirtschaftsmathematik, 1998. submitted.

[MT82] N. Megiddo and A. Tamir. On the complexity of locating linear facilities in the plane. *Operations Research Letters*, 1(5):194–197, 1982.

[MT83] N. Megiddo and A. Tamir. Finding least-distance lines. *SIAM J. on Algebraic and Discrete Methods*, 4(2):207–211, 1983.

[Nic95] S. Nickel. *Discretization of Planar Location Problems.* Shaker Verlag, Aachen, 1995.

[Nie96] F. Nielsen. Fast stabbing of boxes in high dimensions. In *Proceedings of the 8th Canadian Conference on Computational Geometry,* pages 87–92, Ottawa, Canada, 1996. Carleton University Press.

[NM80] J.P. Norback and J.G. Morris. Fitting hyperplanes by minimizing orthogonal deviations. *Math. Programming,* 19:102–105, 1980.

[NS97] S. Nickel and A. Schöbel. A geometric approach to global optimization. Technical Report 27, University of Kaiserslautern, Wirtschaftsmathematik, 1997. to appear in Journal of Global Optimization.

[NW88] G.L. Nemhauser and L.A. Wolsey. *Integer and Combinatorial Optimization.* John Wiley & Sons, New York-Chichester-Brisbane-Toronto-Singapore, 1988.

[OvL81] M.H. Overmars and J. van Leeuwen. Dynamically maintaining configurations in the plane. *Journal Comput. Syst. Sci,* 23:166–204, 1981.

[Pee96] P. Peeters. A counterexample for the circle location problem. personal communication, 1996.

[Pet97] T.J. Petersen. The 1-segment center problem. Master's thesis, The University of Copenhagen, 1997.

[PFTV86] W.H. Press, B.P. Flannery, S.A. Teulosky, and W.T. Vetterling. *Numerical Recipes.* Cambridge University Press, Cambridge, New York, 1986.

[Phe89] R.R. Phelps. *Convex Functions, Monotone Operators and Differentiability,* volume 1364 of *Lecture Notes in Mathematics.* Springer, Berlin, 1989.

[Pla95] F. Plastria. Continuous location problems. In Z. Drezner, editor, *Facility Location: A survey of applications and methods,* chapter 11, pages 225–262. Springer, New York, Inc., 1995.

[PSS92] J. Pach, W. Steiger, and E. Szemeredi. An upper bound on the number of planar k–sets. *Discrete Comput. Geometry,* 7:109–123, 1992.

[PW90] R. Pollack and R. Wenger. Necessary and sufficient conditions for hyperplane transversals. *Combinatorica,* 10:307–311, 1990.

[Ric64] J. Rice. *The Approximation of Functions: The Linear Theory.* Addison-Wesley, 1964. Volume 1.

[Rob91] J.-M. Robert. *Linear Approximation and Line Transversals.* PhD thesis, School of Computer Sciences, McGill University, Montreal, 1991.

[RS72a] M.R. Rao and V. Srinivasan. A note on Sharpe's algorithm for minimizing the sum of absolute deviations in a simple regression problem. *Management Science*, 19:222–225, 1972.

[RS72b] C.P. Rourke and B.J. Sanderson. *Introduction to Piecewise-Linear Topology.* Springer-Verlag, 1972.

[SA95] M. Sharir and P.K. Agarwal. *Davenport-Schinzel Sequences and their Geometric Applications.* Cambridge University Press, Cambridge, New York, Melbourne, 1995.

[Sch73] E.J. Schlossmacher. An iterative technique for absolute deviations curve fitting. *Journal Amer. Statist. Association*, 68:857–859, 1973.

[Sch96] A. Schöbel. Locating least-distant lines with block norms. *Studies in Locational Analysis*, 10:139–150, 1996.

[Sch97] A. Schöbel. Locating line segments with vertical distances. *Studies in Locational Analysis*, 11:143–158, 1997.

[Sch98a] A. Schöbel. Locating least distant lines in the plane. *European Journal of Operational Research*, 106(1):152–159, 1998.

[Sch98b] A. Schöbel. Solving restricted line location problems via a dual interpretation. Technical Report 32, University of Kaiserslautern, Wirtschaftsmathematik, 1998. to appear in Discrete Applied Mathematics.

[Sha59] W.G. Sharpe. Mean-absolute deviation characteristic lines for securities and portfolios. *Management Science*, 54:206–212, 1959.

[Sha78] M.I. Shamos. *Computational Geometry.* PhD thesis, Department of Computer Science, Yale University, New Haven, 1978.

[Spä97a] H. Späth. Least squares fitting of ellipses and hyperbolas. *Computational Statistics*, 12(3):329–341, 1997.

[Spä97b] H. Späth. Orthogonal distance fitting by circles and ellipses. *Computational Statistics*, 12(3):343–354, 1997.

[Ste89] R.E. Steuer. *Multiple Criteria Optimization: Theory, Computation, and Application.* Krieger Publishing Company, Malabar, Florida, 1989.

[SW87] H. Späth and G.A. Watson. On orthogonal linear L_1 approximation. *Numer. Math.*, 51:531–543, 1987.

[SW96] M. Sharir and E. Welzl. Rectilinear and polygonal p-piercing and p-center problems. In *Proceedings of the 12th Annual ACM Symposium on Computational Geometry*, pages 122–132, 1996.

[SW98] A. Schöbel and M.M. Wiecek. Bicriteria segment location problems. working paper, 1998.

[Tou83] G.T. Toussaint. Solving geometric problems with rotating calipers. In *Proceedings of the MELECON'83*, Athens, Greece, 1983.

[VY89] J.A. Ventura and S. Yeralan. The minmax center estimation problem. *European Journal of Operational Research*, 41:64–72, 1989.

[Wag59] H.M. Wagner. Linear programming techniques for regression analysis. *Journal Amer. Statist. Association*, 54:206–212, 1959.

[Wen97] R. Wenger. Progress in geometric transversal theory. In *Proceedings of the Mount Holyoke Conference on Discrete and Computational Geometry*, 1997. to appear.

[WGHS97] C. Witzgall, S.I. Gass, H.H. Harary, and D.R. Shier. Linear programming techniques for fitting circles and spheres. Talk held at the ISMP'97, Lausanne, august 1997.

[WW85] J.E. Ward and R.E. Wendell. Using block norms for location modeling. *Operations Research*, 33:1074–1090, 1985.

[YKII88] P. Yamamoto, K. Kato, K. Imai, and H. Imai. Algorithms for vertical and orthogonal L_1 linear approximation of points. Proc. 4th Ann. Sympos. Comput. Geom., pages 352–361, 1988.

[YV88] S. Yeralan and J.A. Ventura. Computerized roundness inspection. *International Journal of Production Research*, 26:1921–1935, 1988.

[Zem84] E. Zemel. An $O(n)$ algorithm for the linear multiple choice knapsack problem and related problems. *Inform. Process. Letters*, 18:123–128, 1984.

[ZV92] R. Zivaljević and S. Vrećica. The colored Tverberg's problem and complexes of injective functions. *Journal Combin. Theory Series A*, 61:309–318, 1992.

Index

MIX
Papier aus verantwortungsvollen Quellen
Paper from responsible sources
FSC® C105338

If you have any concerns about our products,
you can contact us on
ProductSafety@springernature.com

In case Publisher is established outside the EU,
the EU authorized representative is:
Springer Nature Customer Service Center GmbH
Europaplatz 3, 69115 Heidelberg, Germany

Printed by Libri Plureos GmbH
in Hamburg, Germany